SUBSYNCHRONOUS RESONANCE IN POWER SYSTEMS

OTHER IEEE PRESS BOOKS

Teleconferencing, *Edited by D. Bodson and R. Schaphorst*
Polysilicon Emitter Bipolar Transistors, *Edited by A. K. Kapoor and D. J. Roulston*
Integration of Information Systems: Bridging Heterogeneous Databases, *Edited by A. Gupta*
Numerical Methods for Passive Microwave and Millimeter Wave Structures, *Edited by R. Sorrentino*
Visual Communications Systems, *Edited by A. N. Netravali and B. Prasada*
Analog MOS Integrated Circuits, II, *Edited by P. R. Gray, B. A. Wooley, and R. W. Brodersen*
Electrostatic Discharge and Electronic Equipment, *By W. Boxleitner*
Instrumentation and Techniques for Radio Astronomy, *Edited by P. F. Goldsmith*
Network Interconnection and Protocol Conversion, *Edited by P. E. Green, Jr.*
VLSI Signal Processing, III, *Edited by R. W. Brodersen and H. S. Moscovitz*
Microcomputer-Based Expert Systems, *Edited by A. Gupta and B. E. Prasad*
Principles of Expert Systems, *Edited by A. Gupta and B. E. Prasad*
High Voltage Integrated Circuits, *Edited by B. J. Baliga*
Microwave Digital Radio, *Edited by L. J. Greenstein and M. Shafi*
Oliver Heaviside: Sage in Solitude, *By P. J. Nahin*
Radar Applications, *Edited by M. I. Skolnik*
Principles of Computerized Tomographic Imaging, *By A. C. Kak and M. Slaney*
Selected Papers on Noise in Circuits and Systems, *Edited by M. S. Gupta*
Spaceborne Radar Remote Sensing: Applications and Techniques, *By C. Elachi*
Engineering Excellence, *Edited by D. Christiansen*

A complete listing of IEEE PRESS books is available upon request.

SUBSYNCHRONOUS RESONANCE IN POWER SYSTEMS

P. M. Anderson
President and Principal Engineer
Power Math Associates, Inc.

B. L. Agrawal
Senior Consulting Engineer
Arizona Public Service Co.

J. E. Van Ness
Professor of Electrical Engineering and Computer Science
Northwestern University

Published under the sponsorship of the IEEE Power Engineering Society.

IEEE PRESS

The Institute of Electrical and Electronics Engineers, Inc., New York

IEEE Order Number: PP2477

The Library of Congress has catalogued the hard cover edition of this title as follows:

Anderson, P. M. (Paul M.), 1926–
Subsynchronous resonance in power systems/P. M. Anderson, B. L. Agrawal, J. E. Van Ness.
p. cm.
"Published under the sponsorship of the IEEE Power Engineering Society."
Includes bibliographical references.
ISBN 0-87942-258-0 ISBN 0-78035-350-1
1. Electric power system stability—Mathematical models.
2. Subsynchronous resonance (Electrical engineering)—Mathematical models. I. Agrawal, B. L. (Bajarang L.), 1947– . II. Van Ness, J. E. (James E.) III. Title.
TK1005.A73 1989
621.3—dc20 89-28366
CIP

Dedicated to Our Colleagues

Richard G. Farmer

and

Eli Katz

who provided the opportunity for preparation of this book

and gave generously of their special technical knowledge

of Subsynchronous Resonance

TABLE OF CONTENTS

Preface

This book is intended to provide the engineer with technical information on subsynchronous resonance (SSR), and to show how the computation of eigenvalues for the study of SSR in an interconnected power system can be accomplished. It is primarily a book on mathematical modeling. It describes and explains the differential equations of the power system that are required for the study of SSR. However, the objective of modeling is analysis. The analysis of SSR may be performed in several different ways, depending on the magnitude of the disturbance and the purpose of the study. The goal here is to examine the *small disturbance* behavior of a system in which SSR oscillations may exist. Therefore, we present the equations to compute the *eigenvalues* of the power system so that the interaction between the network and the turbine-generator units can be studied. Eigenvalue analysis requires that the system be linear. Since turbine-generator equations are nonlinear, the linearization of these equations is also explained in detail. The equations are also normalized to ease the problem of providing data for existing systems and for estimating data for future systems that are under study.

There are many references that describe SSR phenomena, some general or introductory in nature, and others very technical and detailed. The authors have been motivated to provide a book that is tutorial on the subject of SSR, and to provide more detail in the explanations than one generally finds in the technical literature. It is assumed that the user of this book is acquainted with power systems and the general way in which power systems are modeled for analysis. Normalization of the power system equations is performed here, but without detailed explanation. This implies that background study may be required by some readers, and this study is certainly recommended. In some cases, the background reading may be very important. Numerous references are cited to point the way and certain references are mentioned in the text that are believed to be helpful.

The authors wish to acknowledge the support of the Los Angeles Department of Water and Power (DWP) and the Arizona Public Service Company (APS) for sponsoring the work that led to the writing of this book. In particular, the advice and assistance of Eli Katz and Richard Lee of DWP and of Richard Farmer of APS are acknowledged. Mr. Katz was the prime mover in having this work undertaken, and he did so in anticipation of his retirement, at which time he realized that he was about the only person in his company with experience in solving SSR problems. He and Mr. Lee felt that a tutorial reference book would be helpful to their younger colleagues, since there are no textbooks on the subject, and requested that a tutorial report be submitted on the subject. They also felt that their company needed the eigenvalue computation capability to reinforce other methods then in use by their company for SSR studies.

Mr. Farmer of APS also became involved in the project and assisted greatly in its success, drawing on his personal knowledge of the subject. He provided valuable insight and was responsible for focusing our work at the microcomputer level. This had not been previously considered, partly because eigenvalue computation is computer intensive and had "always been done" on large computers. In retrospect, this was a great idea, and we all became quite enthusiastic about it.

This project led to a collaboration among the three authors, and indeed led to the writing of this book. Jim Van Ness was our expert on eigenvalue and eigenvector computation. We used the program PALS that he had written earlier for the Bonneville Power Administration as the backbone code for the eigenvalue/eigenvector calculations. Jim was also responsible for the coding of our additions to that backbone program and for testing our equations on his computer to make sure we were getting the right answers.

Baj Agrawal was our expert on many topics, but particularly the specification of data for making SSR studies. His extensive experience in performing system tests to determine these data provided us with valuable insights. We hope that his documentation of this information will be helpful to the reader, especially those who have the responsibility of system testing. Much of this information has never before appeared in a tutorial book before, and is taken from fairly recent research documents.

Paul Anderson provided the material on modeling of the system, its transformation, and normalization. He worked on much of the descriptive material for the book and served as a managing editor to see that it all came together in the same language, if not in the same style.

It was a good collaboration for the three of us and we learned to appreciate the expertise of our colleagues as we worked together. We sincerely hope that this comes through for the reader and that the book might be as interesting for the engineer to read as it was for us to prepare.

The authors would like to thank several individuals who provided valuable assistance in the preparation and checking of the manuscript. Most of the figures were prepared on the computer by Garrett Rusch, a student at the University of California at San Diego, whose skill in computer graphics drafting is acknowledged. We are also indebted to Jai-Soo Jang, a graduate student at Northwestern University, who studied the entire manuscript and found many typographical errors that we were glad to have corrected. We also thank Mahmood Mirheydar for his work in preparing data in a convenient form for plotting. Finally, we extend a special thanks Dr. Christopher Pottle of Cornell University, who helped us to understand the proper methods for modeling the network for eigenvalue calculations and provided us with a computer program for this evaluation.

For those who might be interested in the details of producing a book of this kind, a few facts concerning its production may be of interest. This book was written entirely on a Macintosh®[1] computer using the program Word® 4.0[2]. All the line drawings were produced using MacDraw® and MacDraw®II[3], and the plots were produced using the Igor©[4] program. All equations were written using the program Mathtype®[5] . The pages were printed using a Linotronic®[6] 300 printer, at a resolution of 1270 dots per inch. The typeface is New Century Schoolbook, and was chosen for its clarity and style, and because it lends itself well to mathematical expressions. The personal computer process permitted the authors to deliver camera ready copy directly to IEEE. Since the text did not have to be reset by a professional typographer, the usual process of page proofs and galleys was thereby eliminated. This saved a great deal of time and prevented the introduction of errors in the retyping of the entire book and, especially, the equations. This is the first book published by IEEE using this process, but will surely not be the last.

P. M. Anderson
B. L. Agrawal
J. E. Van Ness

---

[1]Macintosh is a registered trademark of Apple Computer, Inc.
[2]Microsoft Word is a registered trademark of Microsoft.
[3]MacDraw and MacDraw II are registered trademarks of Claris Corporation.
[4]Igor is a registered trademark of WaveMetrics
[5]Mathtype is a registered trademark of Design Science, Inc.
[6]Linotronic is a registered trademark of Linotype AG.

SUBSYNCHRONOUS RESONANCE IN POWER SYSTEMS

Part 1

Introduction

CHAPTER 1

INTRODUCTION

This book provides a tutorial description of the mathematical models and equation formulations that are required for the study of a special class of dynamic power system problems, namely *subsynchronous resonance* (SSR). Systems that experience SSR exhibit dynamic oscillations at frequencies below the normal system base frequency (60 Hz in North America). These problems are of great interest in utilities where this phenomenon is a problem, and the computation of conditions that excite these SSR oscillations are important to those who design and operate these power systems.

This book presents the mathematical modeling of the power system, which is explained in considerable detail. The data that are required to support the mathematical models are discussed, with special emphasis on field testing to determine the needed data. However, the purpose of modeling is to support mathematical analysis of the power system. Here, we are interested in the oscillatory behavior of the system, and the damping of these oscillations. A convenient method of analysis to determine this damping is to compute the eigenvalues of a linear model of the system. Eigenvalues that have negative real parts are damped, but those with positive real parts represent resonant conditions that can lead to catastrophic results. Therefore, the computation of eigenvalues and eigenvectors for the study of SSR is an excellent method of providing crucial information about the nature of the power system. The method for computing eigenvalues and eigenvectors is presented, and the interpretation of the resulting information is described.

1.1 DEFINITION OF SSR

Subsynchronous resonance (SSR) is a dynamic phenomenon of interest in power systems that have certain special characteristics. The formal definition of SSR is provided by the IEEE [1]:

> *Subsynchronous resonance* is an electric power system condition where the electric network exchanges energy with a turbine generator at one or more of the natural frequencies of the combined system below the synchronous frequency of the system.

The definition includes any system condition that provides the opportunity for an exchange of energy at a given subsynchronous frequency. This includes what might be considered "natural" modes of oscillation that are due to the inherent system characteristics, as well as "forced" modes of oscillation that are driven by a particular device or control system.

The most common example of the natural mode of subsynchronous oscillation is due to networks that include series capacitor compensated transmission lines. These lines, with their series LC combinations, have natural frequencies ω_n that are defined by the equation

$$\omega_n = \sqrt{\frac{1}{LC}} = \omega_B \sqrt{\frac{X_C}{X_L}} \tag{1.1}$$

where ω_n is the natural frequency associated with a particular line LC product, ω_B is the system base frequency, and X_L and X_C are the inductive and capacitive reactances, respectively. These frequencies appear to the generator rotor as modulations of the base frequency, giving both subsynchronous and supersynchronous rotor frequencies. It is the subsynchronous frequency that may interact with one of the natural torsional modes of the turbine-generator shaft, thereby setting up the conditions for an exchange of energy at a subsynchronous frequency, with possible torsional fatigue damage to the turbine-generator shaft.

The torsional modes (frequencies) of shaft oscillation are usually known, or may be obtained from the turbine-generator manufacturer. The network frequencies depend on many factors, such as the amount of series capacitance in service and the network switching arrangement at a particular time. The engineer needs a method for examining a large number of feasible operating conditions to determine the possibility of SSR interactions. The eigenvalue program provides this tool. Moreover, the eigenvalue computation permits the engineer to track the locus of system eigenvalues as parameters such as the series capacitance are varied to represent equipment outages. If the locus of a particular eigenvalue approaches or crosses the imaginary axis, then a critical condition is identified that will require the application of one or more SSR countermeasures [2].

1.2 POWER SYSTEM MODELING

This section presents an overview of power system modeling and defines the limits of modeling for the analysis of SSR. We are interested here in modeling the power system for the study of dynamic performance. This means that the system is described by a system of differential equations.

Usually, these equations are nonlinear, and the complete description of the power system may require a very large number of equations. For example, consider the interconnected network of the western United States, from the Rockies to the Pacific, and the associated generating sources and loads. This network consists of over 3000 buses and about 400 generating stations, and service is provided to about 800 load points. Let us assume that the network and loads may be defined by algebraic models for the analytical purpose at hand. Moreover, suppose that the generating stations can be modeled by a set of about 20 first order differential equations. Such a specification, which might be typical of a transient stability analysis, would require 8000 differential equations and about 3500 algebraic equations. A very large number of oscillatory modes will be present in the solution. This makes it difficult to understand the effects due to given causes because so many detailed interactions are represented.

Power system models are often conveniently defined in terms of the major subsystems of equipment that are active in determining the system performance. Figure 1.1 shows a broad overview of the bulk power system, including the network, the loads, the generation sources, the system control, the telecommunications, and the interconnections with neighboring utilities. For SSR studies we are interested in the prime mover (turbines) and generators and their primary controls, the speed governors and excitation systems. The network is very important and is represented in detail, but using only algebraic equations and ordinary differential equations (lumped parameters) rather than the exact partial differential equations. This is because we are interested only in the low frequency performance of the network, not in traveling waves. The loads may be important, but are usually represented as constant impedances in SSR modeling. We are not interested in the energy sources, such as boilers or nuclear reactors, nor are we concerned about the system control center, which deals with very low frequency phenomena, such as daily load tracking. These frequencies are too low for concern here.

Clearly, the transient behavior of the system ranges from the dynamics of lightning surges to that of generation dispatch and load following, and covers several decades of the frequency domain, as shown in Figure 1.2. Note that SSR falls largely in the middle of the range depicted, with major emphasis in the subsynchronous range. Usually, we say that the frequencies of oscillation that are of greatest interest are those between about 10 and 50 Hz. We must model frequencies outside of this narrow band, however, since modulations of other interactions may produce frequencies in the band of interest. It is noted, from Figure 1.2, that the.

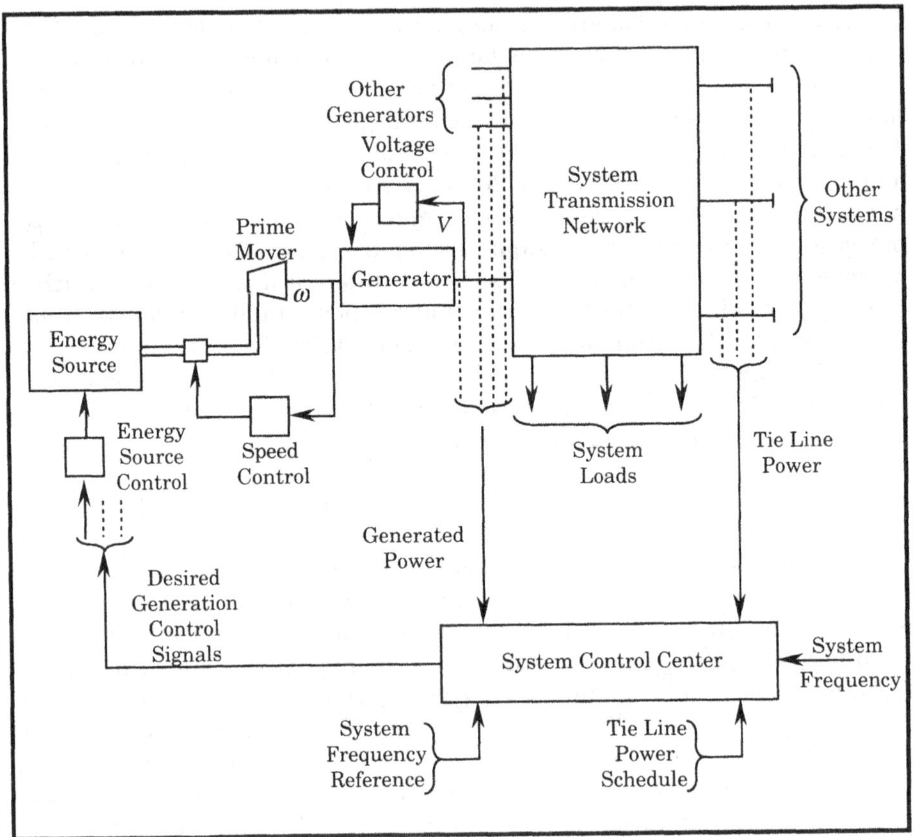

Figure 1.1 Structure of a Power System for Dynamic Analysis

basic range of frequencies of interest is not greatly different from transient stability. Hence, many of the models from transient stability will be appropriate to use.

In modeling the system for analysis, we find it useful to break the entire system up into physical subsystems, as in Figure 1.3, which shows the major subsystems associated with a single generating unit and its interconnection with the network and controls. In SSR analysis, it is necessary to model most, but not all, of these subsystems, and it is necessary to model at least a portion of the network. The subset of the system to be modeled for SSR is labeled in Figure 1.3, where the shaded region is the subset of interest in many studies. Also, it is usually necessary to model several machines for SSR studies, in addition to the interface between each machine and the network.

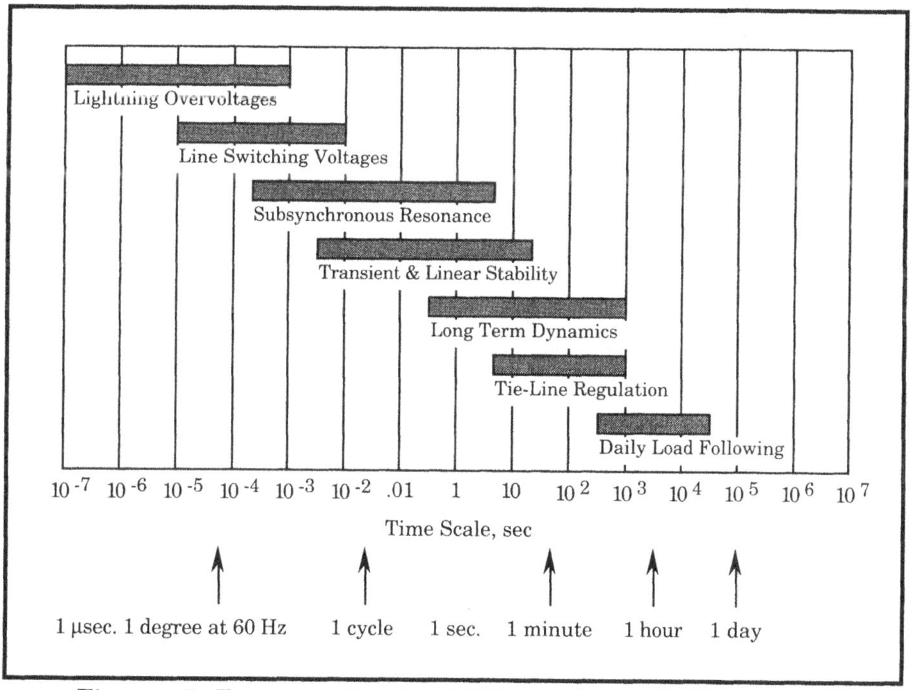

Figure 1.2 Frequency Bands of Different Dynamic Phenomena

Figure 1.3 also shows a convenient definition of the inputs and outputs defined for each subsystem model. The shaded subset defined in this figure is somewhat arbitrary. Some studies may include models of exciters, speed governors, high voltage direct current (HVDC) converter terminals, and other apparatus. It would seldom be necessary to model a boiler or nuclear reactor for SSR studies. The shaded area is that addressed in this book. Extensions of the equations developed for subsystems shown in Figure 1.3 should be straightforward.

In modeling the dynamic system for analysis, one must first define the scope of the analysis to be performed, and from this scope define the modeling limitations. No model is adequate for all possible types of analysis. Thus, for SSR analytical modeling we define the following scope:

> **Scope of SSR Models** The scope of SSR models to be derived in this monograph is limited to the dynamic performance of the interactions between the synchronous machine and the electric network in the subsynchronous frequency range, generally between 0 and 50 Hz. The subsystems defined for modeling are the following:

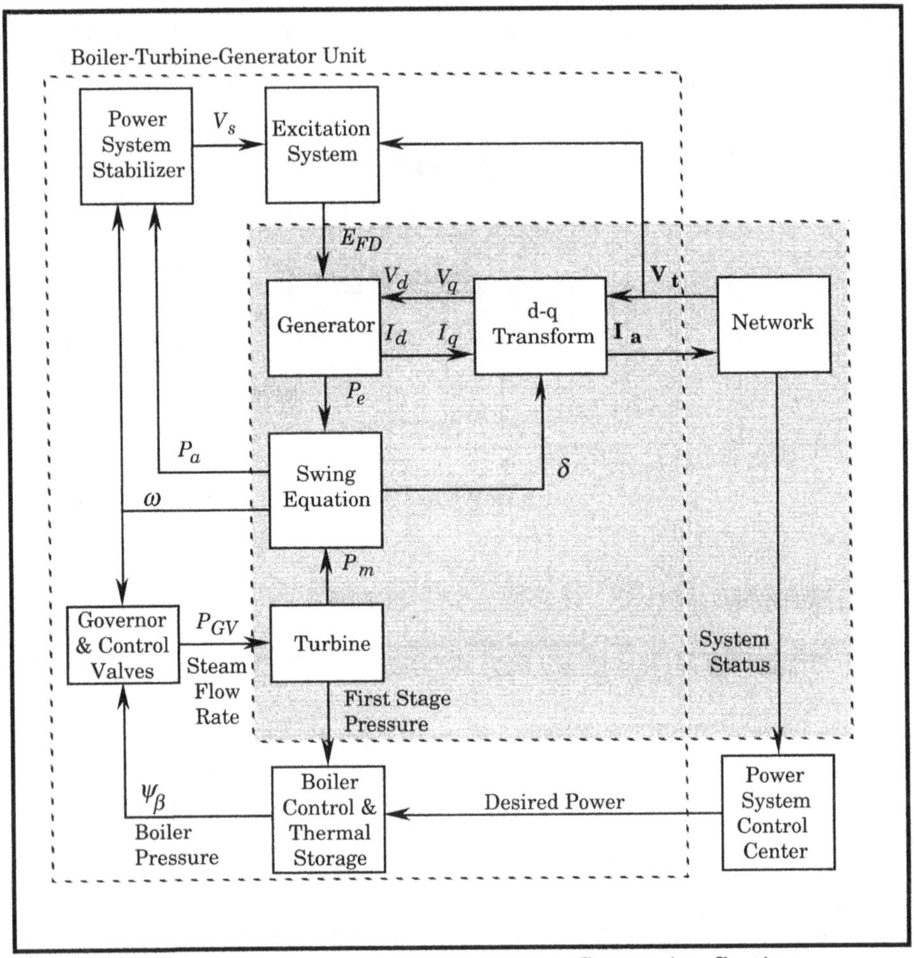

Figure 1.3 Subsystems of Interest at a Generating Station

- Network transmission lines, including series capacitors.
- Network static shunt elements, consisting of R, L, and C branches.
- Synchronous generators.
- Turbine-generator shafts with lumped spring-mass representation and with self and mutual damping.
- Turbine representation in various turbine cylinder configurations.

It is also necessary to define the approximate model bandwidth considered essential for accurate simulated performance of the system under study. For the purpose here, models will be derived that have a bandwidth of about 60 Hz.

1.3 INTRODUCTION TO SSR

Subsynchronous resonance is a condition that can exist on a power system wherein the network has natural frequencies that fall below the nominal 60 hertz of the network applied voltages. Currents flowing in the ac network have two components; one component at the frequency of the driving voltages (60 Hz) and another sinusoidal component at a frequency that depends entirely on the elements of the network. We can write a general expression for the current in a simple series R-L-C network as

$$i(t) = K\left[A\sin(\omega_1 t + \psi_1) + Be^{-\varsigma\omega_2 t}\sin(\omega_2 t + \psi_2)\right] \tag{1.2}$$

where all of the parameters in the equation are functions of the network elements except ω_1, which is the frequency of the driving voltages of all the generators. Note that even ω_2 is a function of the network elements.

Currents similar to (1.2) flow in the stator windings of the generator and are reflected into the generator rotor a physical process that is described mathematically by Park's transformation. This transformation makes the 60 hertz component of current appear, as viewed from the rotor, as a dc current in the steady state, but the currents of frequency ω_2 are transformed into currents of frequencies containing the sum $(\omega_1+\omega_2)$ and difference $(\omega_1-\omega_2)$ of the two frequencies. The difference frequencies are called *subsynchronous* frequencies. These subsynchronous currents produce shaft torques on the turbine-generator rotor that cause the rotor to oscillate at subsynchronous frequencies.

The presence of subsynchronous torques on the rotor causes concern because the turbine-generator shaft itself has natural modes of oscillation that are typical of any spring mass system. It happens that the shaft oscillatory modes are at subsynchronous frequencies. Should the induced subsynchronous torques coincide with one of the shaft natural modes of oscillation, the shaft will oscillate at this natural frequency, sometimes with high amplitude. This is called subsynchronous resonance, which can cause shaft fatigue and possible damage or failure.

1.3.1 Types of SSR Interactions

There are many ways in which the system and the generator may interact with subsynchronous effects. A few of these interactions are basic in concept and have been given special names. We mention three of these that are of particular interest:

Induction Generator Effect
Torsional Interaction Effect
Transient Torque Effect

Induction Generator Effect

Induction generator effect is caused by self excitation of the electrical system. The resistance of the rotor to subsynchronous current, viewed from the armature terminals, is a negative resistance. The network also presents a resistance to these same currents that is positive. However, if the negative resistance of the generator is greater in magnitude than the positive resistance of the network at the system natural frequencies, there will be sustained subsynchronous currents. This is the condition known as the "induction generator effect."

Torsional Interaction

Torsional interaction occurs when the induced subsynchronous torque in the generator is close to one of the torsional natural modes of the turbine-generator shaft. When this happens, generator rotor oscillations will build up and this motion will induce armature voltage components at both subsynchronous and supersynchronous frequencies. Moreover, the induced subsynchronous frequency voltage is phased to sustain the subsynchronous torque. If this torque equals or exceeds the inherent mechanical damping of the rotating system, the system will become self-excited. This phenomenon is called "torsional interaction."

Transient Torques

Transient torques are those that result from system disturbances. System disturbances cause sudden changes in the network, resulting in sudden changes in currents that will tend to oscillate at the natural frequencies of the network. In a transmission system without series capacitors, these transients are always dc transients, which decay to zero with a time constant that depends on the ratio of inductance to resistance. For networks that contain series capacitors, the transient currents will be of a form similar to equation (1.2), and will contain one or more oscillatory frequencies that depend on the network capacitance as well as the inductance and resistance. In a simple radial R-L-C system, there will be only one such natural frequency, which is exactly the situation described in

(1.2), but in a network with many series capacitors there will be many such subsynchronous frequencies. If any of these subsynchronous network frequencies coincide with one of the natural modes of a turbine-generator shaft, there can be peak torques that are quite large since these torques are directly proportional to the magnitude of the oscillating current. Currents due to short circuits, therefore, can produce very large shaft torques both when the fault is applied and also when it is cleared. In a real power system there may be many different subsynchronous frequencies involved and the analysis is quite complex.

Of the three different types of interactions described above, the first two may be considered as small disturbance conditions, at least initially. The third type is definitely not a small disturbance and nonlinearities of the system also enter into the analysis. From the viewpoint of system analysis, it is important to note that the induction generator and torsional interaction effects may be analyzed using linear models, suggesting that eigenvalue analysis is appropriate for the study of these problems.

1.3.2 Analytical Tools

There are several analytical tools that have evolved for the study of SSR. The most common of these tools will be described briefly.

Frequency Scanning

Frequency scanning is a technique that has been widely used in North America for at least a preliminary analysis of SSR problems, and is particularly effective in the study of induction generator effects. The frequency scan technique computes the equivalent resistance and inductance, seen looking into the network from a point behind the stator winding of a particular generator, as a function of frequency. Should there be a frequency at which the inductance is zero and the resistance negative, self sustaining oscillations at that frequency would be expected due to induction generator effect.

The frequency scan method also provides information regarding possible problems with torsional interaction and transient torques. Torsional interaction or transient torque problems might be expected to occur if there is a network series resonance or a reactance minimum that is very close to one of the shaft torsional frequencies.

Figure 1.4 shows the plot of a typical result from a frequency scan of a network [3]. The scan covers the frequency range from 20 to 50 hertz and shows separate plots for the resistance and reactance as a function of

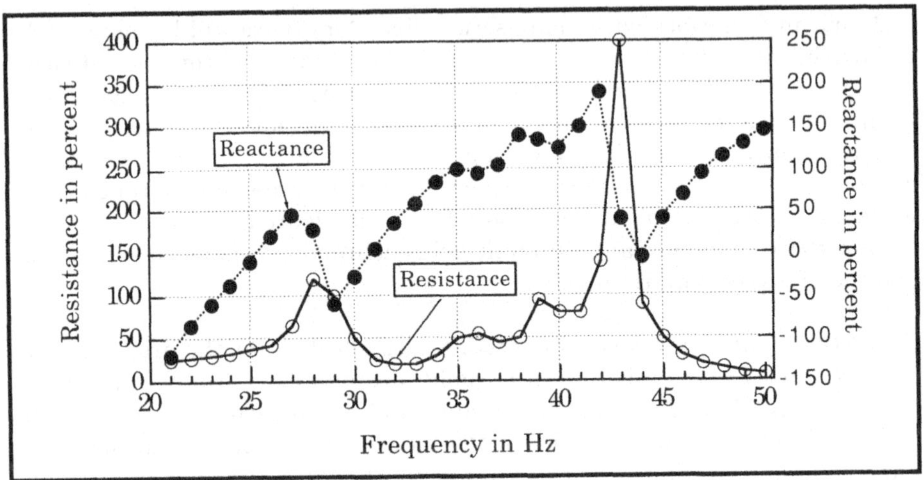

Figure 1.4 Plot from the Frequency Scan of a Network [3]

frequency. The frequency scan shown in the figure was computed for a generator connected to a network with series compensated transmission lines and represents the impedance seen looking into that network from the generator. The computation indicates that there may be a problem with torsional interactions at the first torsional mode, which occurs for this generator at about 44 Hz. At this frequency, the reactance of the network goes to zero, indicating a possible problem. Since the frequency scan results change with different system conditions and with the number of generators on line, many conditions need to be tested. The potential problem noted in the figure was confirmed by other tests and remedial countermeasures were prescribed to alleviate the problem [3].

Frequency scanning is limited to the impedances seen at a particular point in the network, usually behind the stator windings of a generator. The process must be repeated for different system (switching) conditions at the terminals of each generator of interest.

Eigenvalue Analysis

Eigenvalue analysis provides additional information regarding the system performance. This type of analysis is performed with the network and the generators modeled in one linear system of differential equations. The results give both the frequencies of oscillation as well as the damping of each frequency.

Eigenvalues are defined in terms of the system linear equations, that are written in the following standard form.

Table 1.1 Computed Eigenvalues for the First Benchmark Model

Eigenvalue Number	Real Part, s^{-1}	Imaginary Part, rad/s	Imaginary Part, Hz
1,2	+0.07854636	±127.15560200	±20.2374426
3,4	+0.07818368	±99.70883066	±15.86915327
5,6	+0.04089805	±160.38986053	±25.52683912
7,8	+0.00232994	±202.86306822	±32.28666008
9,10	−0.00000048	±298.17672924	±47.45630037
11	−0.77576318		
12	−0.94796049		
13,14	−1.21804111	±10.59514740	±96.61615878
15,16	−5.54108044	±136.97740321	±21.80063081
17,18	−6.80964255	±616.53245850	±98.12275595
19	−25.41118956		
20	−41.29551248		

$$\dot{\mathbf{x}} = \mathbf{A}\mathbf{x} + \mathbf{B}\mathbf{u} \tag{1.3}$$

Then the eigenvalues are defined as the solutions to the matrix equation

$$\det[\lambda \mathbf{U} - \mathbf{A}] = 0 \tag{1.4}$$

where the parameters λ are called the eigenvalues.

An example of eigenvalue analysis is presented using the data from the First Benchmark Model, a one machine system used for SSR program testing [4]. The results of the eigenvalue calculation is shown in Table 1.1. Note that this small system is of 20th order and there are 10 eigenvalues in the range of 15.87 to 47.46 Hz, which is the range where torsional interaction usually occur. Moreover, eight of the eigenvalues have positive real parts, indicating an absence of damping in these modes of response.

Eigenvalue analysis is attractive since it provides the frequencies and the damping at each frequency for the entire system in a single calculation.

EMTP Analysis

The ElectroMagnetic Transients Program (EMTP) is a program for numerical integration of the system differential equations. Unlike a transient stability program, which usually models only positive sequence quantities representing a perfectly balanced system, EMTP is a full three-phase model of the system with much more detailed models of transmission lines, cables, machines, and special devices such as series capacitors with complex bypass switching arrangements. Moreover, the EMTP permits nonlinear modeling of complex system components. It is, therefore, well suited for analyzing the transient torque SSR problems.

The full scope of modeling and simulation of systems using EMTP is beyond the scope of this book. However, to illustrate the type of results that can be obtained using this method, we present one brief example. Figure 1.5 shows the torque at one turbine shaft section for two different levels of series transmission compensation, a small level of compensation for Case A and a larger level for Case B [5]. The disturbance is a three phase fault at time $t = 0$ that persists for 0.06 seconds. It is apparent that the Case B, the higher level of series compensation, results is considerably torque amplification. This type of information would not be available from a frequency scan or from eigenvalue computation, although those methods would indicate the existence of a resonant condition at the indicated frequency of oscillation. EMTP adds important data on the magnitude of the oscillations as well as their damping.

Summary

Three prominent methods of SSR analysis have been briefly described. Frequency scanning provides information regarding the impedance seen, as a function of frequency, looking into the network from the stator of a generator. The method is fast and easy to use. Eigenvalue analysis provides a closed form solution of the entire network including the machines. This gives all of the frequencies of oscillation as well as the damping of each frequency. The method requires more modeling and data than frequency scanning and requires greater computer resources for the computation. EMTP requires still greater modeling effort and computer resources, but allows the full nonlinear modeling of the system machines and other devices, such as capacitor bypass schemes.

In the balance of this book, we concentrate only on the eigenvalue method of SSR analysis. Most of the book is devoted to the mathematical modeling and the determination of accurate model parameters for eigenvalue analysis. First, however, we discuss briefly the types of models used for the SSR

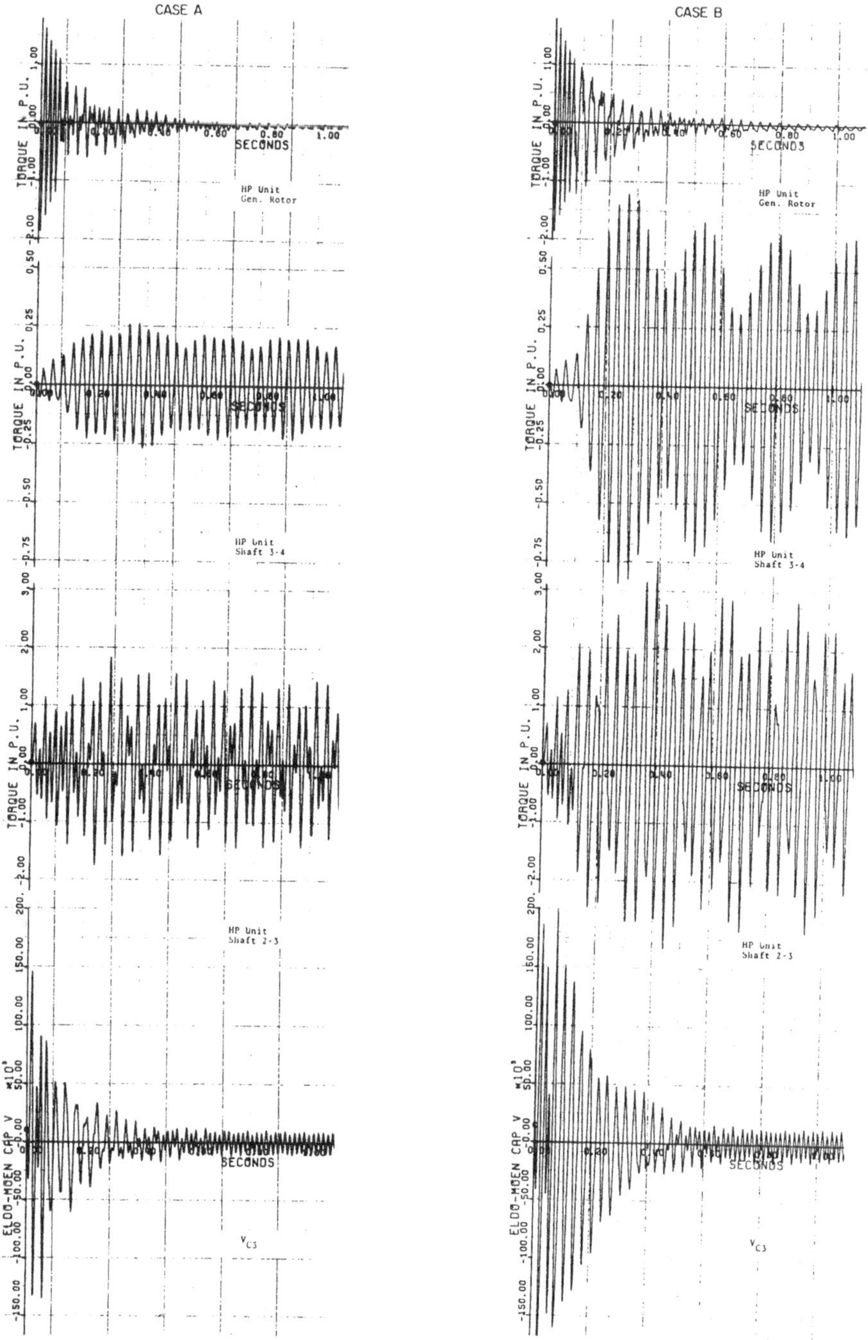

Figure 1.5 Typical Computed Generator Shaft Torques (upper 3 traces) and Voltage Across a Series Capacitor (bottom trace) Using EMTP [5]

analysis. Then we comment briefly on the computed results and their use by the system analyst. Finally, we conclude this chapter with some results from an actual system study to illustrate the way in which eigenvalue calculations may be used.

1.4 EIGENVALUE ANALYSIS

Eigenvalue analysis uses the standard linear, state-space form of system equations and provides an appropriate tool for evaluating system conditions for the study of SSR, particularly for induction generator and torsional interaction effects.

1.4.1 Advantages of Eigenvalue Computation

The advantages of eigenvalue analysis are many. Some of the prominent advantages are:

- Uses the state-space equations, making it possible to utilize many other analytical tools that use this same equation form
- Compute all the exact modes of system oscillation in a single computation
- Can be arranged to perform a convenient parameter variation to study parameter sensitivities
- Can be used to plot root loci of eigenvalue movement in response to many different types of changes

Eigenvalue analysis also includes the computation of *eigenvectors*, which are often not as well understood as eigenvalues, but are very important quantities for analyzing the system. Very briefly, there are two types of eigenvectors, usually called the "right hand" and "left hand" eigenvectors. These quantities are used as follows:

- Right Hand Eigenvectors - show the distribution of modes of response (eigenvalues) through the state variables
- Left Hand Eigenvectors - show the relative effect of different initial conditions of the state variables on the modes of response (eigenvalues)

The right hand eigenvectors are the most useful in SSR analysis. Using these vectors, one can establish the relative magnitude of each mode's response due to each state variable. In this way, one can determine those state variables that have little or no effect on a given mode of response and, conversely, those variables that an play important role is contributing to a given response. This often tells the engineer exactly those variables that need to be controlled in order to damp a subsynchronous oscillation on a given unit.

1.4.2 Disadvantages of Eigenvalue Calculation

Eigenvalue analysis is computationally intensive and is useful only for the linear problem. Moreover, this type of analysis is limited to relatively small systems, say of 500th order or less. Recent work has been done on much larger systems, but most of these methods compute only selected eigenvalues and usually require a skilled and experienced analyst in order to be effective [8,9]. Work is progressing on more general methods of solving large systems, but no breakthroughs have been reported.

Another difficulty of eigenvalue analysis is the general level of difficulty in writing eigenvalue computer programs. Much work has been done in this area, and the SSR analyst can take advantage of this entire realm of effort. Perhaps the most significant work is that performed over the years by the Argonne National Laboratory, which has produced the public domain program known as EISPACK [10]. Another program called PALS has been developed by Van Ness for the Bonneville Power Administration, using some special analytical techniques [11]. Thus, there are complete programs available to those who wish to pursue eigenvalue analysis without the difficult startup task of writing an eigenvalue program.

1.5 CONCLUSIONS

In this chapter, we have reviewed the study of subsynchronous resonance using eigenvalue analysis. From our analysis of the types of SSR interactions, we conclude that eigenvalue analysis is appropriate for the study of induction generator and torsional interaction effects. This will not cover all of the concerns regarding SSR hazards, but it does provide a method of analyzing some of the basic problems.

The system modeling for eigenvalue analysis must be linear. Linear models must be used for the generator, the turbine-generator shaft, and the network. These models are not much different than those used for other types of analysis, except that nonlinearities must be eliminated in the equations. These models are described in Chapters 2, 3, and 4. Another problem related to modeling is the determination of accurate data, either from records of the utility or manufacturer, or from field testing. This important subject is discussed in Chapters 5 and 6.

Eigenvalue and eigenvector computation provide valuable insight into the dynamics of the power system. It is important to identify the possibility of negative damping due to the many system interactions, and the eigenvalue computation does this very clearly. Moreover, eigenvector computation provides a powerful tool to identify those states of the system that lead to various modes of oscillation, giving the engineer a valuable method of designing effective SSR countermeasures. Eigenvalues and eigenvector computations are described more fully in Chapter 7.

Finally, we have illustrated the type of eigenvalue calculation that is performed by showing data from actual system tests to determine damping parameters and the application of these parameters to assure proper damping of various modes of oscillation. The final chapter of the book provides the solution to several "benchmark" problems. These solved cases provide the reader with a convenient way of checking computations made with any eigenvalue program.

1.6 PURPOSE, SCOPE, AND ASSUMPTIONS

The purpose of this monograph is to develop the theory and mathematical modeling of a power system for small disturbance (linear) analysis of subsynchronous resonance phenomena. This theoretical background will provide the necessary linear dynamic equations required for eigenvalue analysis of a power system, with emphasis on the problems associated with SSR. Because the scope is limited to linear analysis of SSR, several important assumptions regarding the application of the system models are necessary. These assumptions are summarized as follows:

1. The turbine-generator initial conditions are computed from a steady-state power flow of the system under study.

2. All system nonlinearities can be initialized and linearized about the initial operating point.

3. The network and loads may be represented as a balanced three-phase system with impedances in each phase equal to the positive sequence impedance.

4. The synchronous generators may be represented by a Park's two-axis model with negligible zero-sequence current.

5. The turbine-generator shaft may be represented as a lumped spring-mass system, with adjacent masses connected by shaft

stiffness and damping elements, and with damping between each mass and the stationary support of the rotating system.

6. Nonlinear controllers may be represented as continuous linear components with appropriately derived linear parameters.

1.7 GUIDELINES FOR USING THIS BOOK

This book is intended as a complete and well documented introduction to the modeling of the major power system elements that are required for SSR analysis. The analytical technique of emphasis is eigenvalue analysis, but many of the principles are equally applicable to other forms of analysis. The major assumption required for eigenvalue analysis is that of linearity, which may make the equations unsuitable for other applications. The nonlinear equations, from which these linear forms are derived, may be necessary for a particular application.

This book does not attempt proofs or extensive derivations of system equations, and the reader must refer to more academic sources for this kind of detailed assistance. Many references to suggested sources of background information are provided. It is assumed that the user of this book is an engineer or scientist with training in the physical and mathematical sciences. These basic study areas are not reviewed or presented in any way, but are used with the assumption that a trained person will be able to follow the developments, probably without referring to other resources.

The major topic of interest here is SSR, and all developments are presented with this objective in mind. We presume that the reader is interested in learning about SSR or wishes to review the background material pertinent to the subject. With this objective foremost, we suggest that the first-time user attempt a straight-through superficial reading of the book in order to obtain an overall grasp of the subject and an understanding of the modeling objectives and interfaces. This understanding should be followed by returning to those sections that require additional study for better understanding or for reinforcing the modeling task at hand.

The second objective of this work is to present a discussion of eigen analysis and to explain the meaning of results that are obtainable from eigenvalue-eigenvector computation. These calculations must be performed by digital computer using very large and complex computer codes. We do not attempt an explanation of these codes or the complex algorithmic development that makes these calculations possible. This area is considered much more detailed than the average engineer would find useful. We do feel, however, that the user should have a sense of what the eigenprogram is used for and should be able to interpret the results of these calculations. In this sense, this document stands as a background reference to the eigenvalue programs [4].

A third objective of this book is to present a discussion of the problems associated with preparing data for use in making SSR eigenvalue-eigenvector calculations. A simulation is of no value whatever if the input data is incorrect or is improperly prepared. Thus it is necessary to understand the modeling and to be able to interpret the data made available by the manufacturers in order to avoid the pitfall of obtaining useless results due to inadequate preparation of study data. This may require the use of judgment, for example, for interpreting the need for a data item that is not immediately available. It may also provide guidance for identifying data that should be obtained by field tests on the actual equipment installed on the system.

1.8 SSR REFERENCES

There are many references on the subjects of concern in this book. This review of prior work is divided into three parts: general references, SSR references, and eigenvalue applications to power systems.

1.8.1 General References

The general references of direct interest in this book are **Power System Control and Stability**, by Anderson and Fouad [14], **Power System Stability, vol 1, 2, and 3**, by Kimbark [15-17], **Stability of Large Electric Power Systems**, by Byerly and Kimbark [18], **The General Theory of Electrical Machines**, by Adkins [19], **The Principles of Synchronous Machines**, by Lewis [20], and **Synchronous Machines**, by Concordia [21].

The material presented in this book is not new and is broadly based on the above references, but with emphasis on the SSR problem.

1.8.2 SSR References

SSR has been the subject of many technical papers, published largely in the past decade. These papers are summarized in three bibliographies [22-24], prepared by the IEEE Working Group on Subsynchronous Resonance (hereafter referred to as the IEEE WG). The IEEE WG has also been responsible for two excellent general references on the subject, which were published as the permanent records of two IEEE Symposia on SSR. The first of these, "Analysis and Control of Subsynchronous Resonance" [25] is largely tutorial and describes the state of the art of the subject. The second document, "Symposium on Countermeasures for Subsynchronous Resonance" [26] describes various approaches used by utilities to analyze and design SSR protective strategies and controls.

In addition to these general references on SSR, the IEEE WG has published six important technical papers on the subject. The first of these, "Proposed Terms and Definitions for Subsynchronous Oscillations" [27] provides an important source for this monograph in clarifying the terminology of the subject area. A later paper, "Terms, Definitions and Symbols for Subsynchronous Oscillations" [28] provides additional definitions and clarifies the original paper. This document is adhered to as a standard in this book. Another IEEE WG report, "First Benchmark Model for Computer Simulation of Subsynchronous Resonance" [4], provides a simple one machine model and test problem for computer program verification and comparison. This was followed by a more complex model described in the paper "Second Benchmark Model for Computer Simulation of Subsynchronous Resonance" [29], which provides a more complex model and test system. A third paper, "Countermeasures to Subsynchronous Resonance Problems" [30], presents a collection of proposed solutions to SSR problems without any attempt at ranking or evaluating the merit of the various approaches. Finally, the IEEE WG published the 1983 prize paper "Series Capacitor Controls and Settings as Countermeasures to Subsynchronous Resonance" [31], which presents the most common system conditions that may lead to large turbine-generator oscillatory torques and describes series capacitor controls and settings that have been successfully applied as countermeasures.

Another publication that contains much information of general importance to the SSR problem is the IEEE document "State-of-the-Art Symposium--Turbine Generator Shaft Torsionals," which describes the problem of stress and fatigue damage in turbine-generator shafts from a variety of causes [32].

1.8.3 Eigenvalue/Eigenvector Analysis References

In the area of eigenvalue analysis there are literally hundreds of papers in the literature. Even those that address power system applications are numerous. We mention here a few references of direct interest. J. H. Wilkinson's book, **The Algebraic Eigenvalue Problem** [12] is a standard reference on the subject. Power system applications can be identified in association with certain authors. We cite particularly the work performed at McMasters University [34-39], that performed at Northwestern University [11, 40-45], the excellent work done at MIT [46], that performed at

Westinghouse[47-49], and the work performed by engineers at Ontario Hydro [50-53]. Also of direct interest is the significant work performed on eigenvalue numerical methods, which resulted in the computer programs known as EISPACK, summarized in [10] and [54].

1.9 REFERENCES FOR CHAPTER 1

1. IEEE SSR Working Group, "Proposed Terms and Definitions for Subsynchronous Resonance," IEEE Symposium on Countermeasures for Subsynchronous Resonance, IEEE Pub. 81TH0086-9-PWR, 1981,p 92-97.

2. IEEE SSR Working Group, "Terms, Definitions, and Symbols for Subsynchronous Oscillations," **IEEE Trans.**, v. PAS-104, June 1985.

3. Farmer, R. G., A. L. Schwalb and Eli Katz, "Navajo Project Report on Subsynchronous Resonance Analysis and Solutions," from the IEEE Symposium Publication **Analysis and Control of Subsynchronous Resonance**, IEEE Pub. 76 CH10660O-PWR

4. IEEE Committee Report, "First Benchmark Model for Computer Simulation of Subsynchronous Resonance," **IEEE Trans.**, v. PAS-96, Sept/Oct 1977, p. 1565-1570.

5. Gross, G., and M. C. Hall, "Synchronous Machine and Torsional Dynamics Simulation in the Computation of Electromagnetic Transients," **IEEE Trans.**, v PAS-97, n 4, July/Aug 1978, p 1074, 1086.

6. Dandeno, P. L., and A. T. Poray, "Development of Detailed Turbogenerator Equivalent Circuits from Standstill Frequency Response Measurements," **IEEE Trans.**, v PAS-100, April 1981, p 1646.

7. Chen, Wai-Kai, **Linear Networks and Systems**, Brooks/Cole Engineering Division, Wadsworth, Belmont, California, 1983.

8. Byerly, R. T., R. J. Bennon and D. E. Sherman, "Eigenvalue Analysis of Synchronizing Power Flow Oscillations in Large Electric Power Systems," **IEEE Trans.**, v PAS-101, n 1, January 1982.

9. Wong, D. Y., G. J. Rogers, B. Porretta and P. Kundur, "Eigenvalue Analysis of Very Large Power Systems," **IEEE Trans.**, v PWRS-3, n 2, May 1988.

10. Smith, B. T., et al., **EISPACK Guide -- Matrix Eigensystem Routines**, Springer-Verlag, New York, 1976.

11. Van Ness, J. E. "The Inverse Iteration Method for Finding Eigenvalues," **IEEE Trans.**, v AC-14, 1969, p 63-66.

12. Wilkinson, J. H. **The Algebraic Eigenvalue Problem**, Oxford University Press, 1965.

13. **SSR/EIGEN User's Manual For The Computation of Eigenvalues and Eigenvectors in Problems Related to Power System Subsynchronous Resonance**, Power Math Associates, Inc., Del Mar California, 1987.

14. Anderson,P. M., and A. A. Fouad, **Power System Control and Stability**, Iowa State University Press, 1977.

15. Kimbark, Edward W., **Power System Stability, v.1, Elements of Stability Calculations**, John Wiley and Sons, New York, 1948.

16. Kimbark, Edward W., **Power System Stability, v.2, Power Circuit Breakers and Protective Relays**, John Wiley and Sons, New York, 1950.

17. Kimbark, Edward W., **Power System Stability, v.3, Synchronous Machines**, John Wiley and Sons, New York, 1950.

18. Byerly, Richard T. and Edward W. Kimbark, **Stability of Large Electric Power Systems**, IEEE Press, IEEE, New York, 1974.

19. Adkins, Bernard, **The General Theory of Electrical Machines**, Chapman and Hall, London, 1964.

20. Lewis, William A., **The Principles of Synchronous Machines**, 3rd Ed., Illinois Institute of Technology Bookstore, 1959.

21. Concordia, Charles, **Synchronous Machines - Theory and Performance**, John Wiley and Sons, New York, 1951.

22. IEEE Committee Report, "A Bibliography for the Study of Subsynchronous Resonance Between Rotating Machines and Power Systems," **IEEE Trans.**, v. PAS-95, n. 1, Jan/Feb 1976, p. 216-218.

23. IEEE Committee Report, "First Supplement to A Bibliography for the Study of Subsynchronous Resonance Between Rotating Machines and Power Systems," ibid, v. PAS-98, n. 6, Nov-Dec 1979, p. 1872-1875.

24. IEEE Committee Report, "Second Supplement to A Bibliography for the Study of Subsynchronous Resonance Between Rotating Machines and Power Systems," ibid, v. PAS-104, Feb 1985, p. 321-327.

25. IEEE Committee Report, "Analysis and Control of Subsynchronous Resonance," IEEE Pub. 76 CH1066-0-PWR, 1976.

26. IEEE Committee Report, "Symposium on Countermeasures for Subsynchronous Resonance, IEEE Pub. 81 TH0086-9-PWR, 1981.

27. IEEE Committee Report, "Proposed Terms and Definitions for Subsynchronous Oscillations," **IEEE Trans.**, v. PAS-99, n. 2, Mar/Apr 1980, p. 506-511.

28. IEEE Committee Report, "Terms, Definitions and Symbols for Subsynchronous Oscillations," ibid, v. PAS-104, June 1985, p. 1326-1334.

29. IEEE Committee Report, "Second Benchmark Model for Computer Simulation of Subsynchronous Resonance," ibid, v PAS-104, May 1985, p 1057-1066.

30. IEEE Committee Report, "Countermeasures to Subsynchronous Resonance," ibid, v. PAS-99, n. 5, Sept/Oct 1980, p. 1810-1817.

31. IEEE Committee Report, "Series Capacitor Controls and Settings as Countermeasures to Subsynchronous Resonance," ibid, v. PAS-101, n. 6, June 1982, p. 1281-1287.

32. IEEE Committee Report, "State-of-the-art Symposium -- Turbine Generator Shaft Torsionals," IEEE Pub. 79TH0059-6-PWR, 1979.

33. Wilkinson, J. H., **The Algebraic Eigenvalue Problem**, Oxford University Press, 1965.

34. Nolan, P. J., N. K. Sinha, and R. T. H. Alden, "Eigenvalue Sensitivities of Power Systems including Network and Shaft Dynamics," **IEEE Trans.,** v. PAS-95, 1976, p. 1318 - 1324.

35. Alden, R. T. H., and H. M. Zein El-Din, "Multi-machine Dynamic Stability Calculations," ibid, v. PAS - 95, 1976, p. 1529-1534.

36. Zein El-Din, H. M. and R. T. H. Alden, "Second-Order Eigenvalue Sensitivities Applied to Power System Dynamics," ibid, v. PAS-96, 1977, p. 1928 - 1935.

37. Zein El-Din, H. M. and R. T. H. Alden, "A computer Based Eigenvalue Approach for Power System Dynamics Stability Calculation," Proc. PICA Conf., May 1977, p. 186-192.

38. Elrazaz, Z., and N. K. Sinha, "Dynamic Stability Analysis of Power Systems for Large Parameter Variations," IEEE paper, PES Summer Meeting, Vancouver, B.C., 1979.

39. Elrazaz, Z., and N. K. Sinha, "Dynamic Stability Analysis for Large Parameter Variations: An Eigenvalue Tracking Approach," IEEE paper A80 088-5, PES Winter Meeting, New York, 1979.

40. Van Ness, J. E., J. M. Boyle, and F. P. Imad, "Sensitivities of Large Multiple-Loop Control Systems," **IEEE Trans.**, v. AC-10, July 1965, p. 308-315.

41. Van Ness, J. E. and W. F. Goddard, "Formation of the Coefficient Matrix of a Large Dynamic System," **IEEE Trans.**, v. PAS-87, Jan 1968, p. 80-83.

42. Pinnello, J. A. and J. E. Van Ness, "Dynamic Response of a Large Power System to a Cycle Load Produced by a Nuclear Accelerator," ibid, v. PAS-90, July/Aug 1971, p. 1856-1862.

43. Van Ness, J. E., F. M. Brasch, Jr., G. L. Landgren, and S.T. Naumann, "Analytical Investigation of Dynamic Instability Occurring at Powerton Station," ibid, v PAS-99, n 4, July/Aug 1980, p 1386-1395.

44. Van Ness, J. E., and F. M. Brasch, Jr., "Polynomial Matrix Based Models of Power System Dynamics," ibid, v. PAS-95, July/Aug 1976, p. 1465-1472.

45. Mugwanya, D. K. and J. E. Van Ness, "Mode Coupling in Power Systems," **IEEE Trans.**, v. PWRS-1, May 1987, p. 264-270.

46. Pérez-Arriaga, I. J., G. C. Verghese, and F. C. Schweppe, "Selective Modal Analysis with Applications to Electric Power Systems, Pt I, Heuristic Introduction, and Pt II, The Dynamic Stability Problem," **IEEE Trans.**,, v. PAS-101, n. 9, September 1982, p. 3117-3134.

47. Bauer, D. L., W. D. Buhr, S. S. Cogswell, D. B. Cory, G. B. Ostroski, and D. A. Swanson, "Simulation of Low Frequency Undamped

Oscillations in Large Power Systems," ibid, v. PAS-94, n. 2, Mar/Apr 1975, p. 207-213.

48. Byerly, R. T., D. E. Sherman, and D. K. McLain, "Normal Modes and Mode Shapes Applied to Dynamic Stability Analysis," ibid, v. PAS-94, n. 2, Mar/Apr 1975, p. 224-229.

49. Busby, E. L., J. D. Hurley, F. W. Keay, and C. Raczkowski, "Dynamic Stability Improvement at Monticello Station -- Analytical Study and Field Test," ibid, v. PAS-98, n. 3, May/June 1979, p. 889-901.

50. Kundur, P. and P. L. Dandeno, "Practical Application of Eigenvalue Techniques in the Analysis of Power Systems Dynamic Stability Problems," 5th Power System Computation Conf., Cambridge, England, Sept. 1975.

51. Kundur, P., D. C. Lee, H. M. Zein-el-Din, "Power System Stabilizers for Thermal Units: Analytical Techniques and On-Site Validation," **IEEE Trans.**, v. PAS-100, 1981, p. 81-95.

52. Lee, D. C., R. E. Beaulieu, and G. J. Rogers," "Effects of Governor Characteristics on Turbo-Generator Shaft Torsionals," ibid, v. PAS-104, 1985, p. 1255-1261.

53. Wong, D. Y., G. J. Rogers, B. Poretta, and P. Kundur, "Eigenvalue Analysis of Very Large Power Systems," ibid, v PWRS-3, 1988, p. 472-480.

54. Garbow, B. S. et al., ed., **EISPACK Guide Extension--Matrix Eigensystem Routines**, Springer-Verlag, New York, 1977.

Part 2

System Modeling

The Generator Model

The Network Model

The Turbine-Generator Shaft Model

CHAPTER 2

THE GENERATOR MODEL

Synchronous machines may be modeled in varying degrees of complexity, depending on the purpose of the model usage. One major difference in machine models is in the complexity assumed for the rotor circuits. This is especially important for solid iron rotors, in which case there are no clearly defined rotor current paths and the rotor flux linkages are difficult to express in terms of simple discrete circuits. For SSR analysis, experience has shown that reasonable results may be obtained by defining two rotor circuits on two different axes that are in space quadrature – the familiar d- and q-axes. This approach will be used in the analysis presented here.

Our procedure will be as follows. First, we will discuss the machine configuration and describe the way a three-phase emf is generated. Then we define the flux linkages of stator and rotor circuits that will completely define the machine circuit performance. Next, we will perform a power invariant transformation that will simplify the stator flux linkage equations. We will then write the voltage equations of the transformed system and simplify the resulting equations for computer analysis.

2.1 THE SYNCHRONOUS MACHINE STRUCTURE

The flux linkage equations for the synchronous machine are defined in terms of the self and mutual inductances of the windings. Figure 2.1 shows an end view of the generator windings, where we have made the following assumptions:

1. The flux density seen by the stator conductors may be considered to be sinusoidal. Actually, a sinusoidal flux density spatial distribution is achieved only approximately in physical machines.

2. The induced emf in each phase can be represented as if produced by an equivalent single coil for that phase, as shown in Figure 2.1. The actual machine has many coils in each phase. Our simple coil representation should be thought of as the net effect of the many phase windings in each phase.

3. Two equivalent rotor circuits are represented in each axis of the rotor – F and D in the d-axis, and G and Q in the q-axis, with positive current direction defined as the direction causing positive magnetization of the defined d- and q-axis direction, respectively.

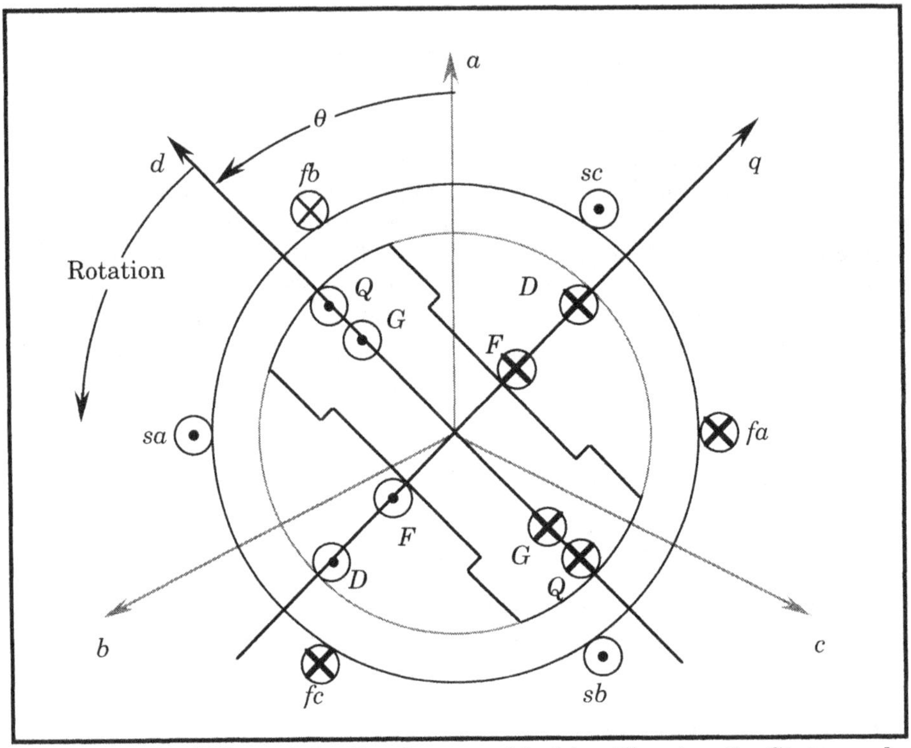

Figure 2.1 End View of the Synchronous Machine Showing the Stator and Rotor Equivalent Coil Locations

4. The positive direction of rotation and the direction of the d- and q-axes are defined in agreement with IEC Standard 34-10 (1975) [1] and IEEE Std. 100-1984 [2].

To understand the action of an ac generator, one should visualize a rotating magnetic flux density wave in the air gap of the machine as shown in Figure 2.2 [4]. This wave links the stator winding, causing each coil of the stator winding to see an alternating flux. This is the mechanism for inducing an alternating voltage. Figure 2.2 shows an approximate picture of this arrangement. The figure is drawn as if the air gap were straight, rather than circular, for simplicity.

We usually assume that the flux density in the air gap has a sinusoidal distribution, which we may write as

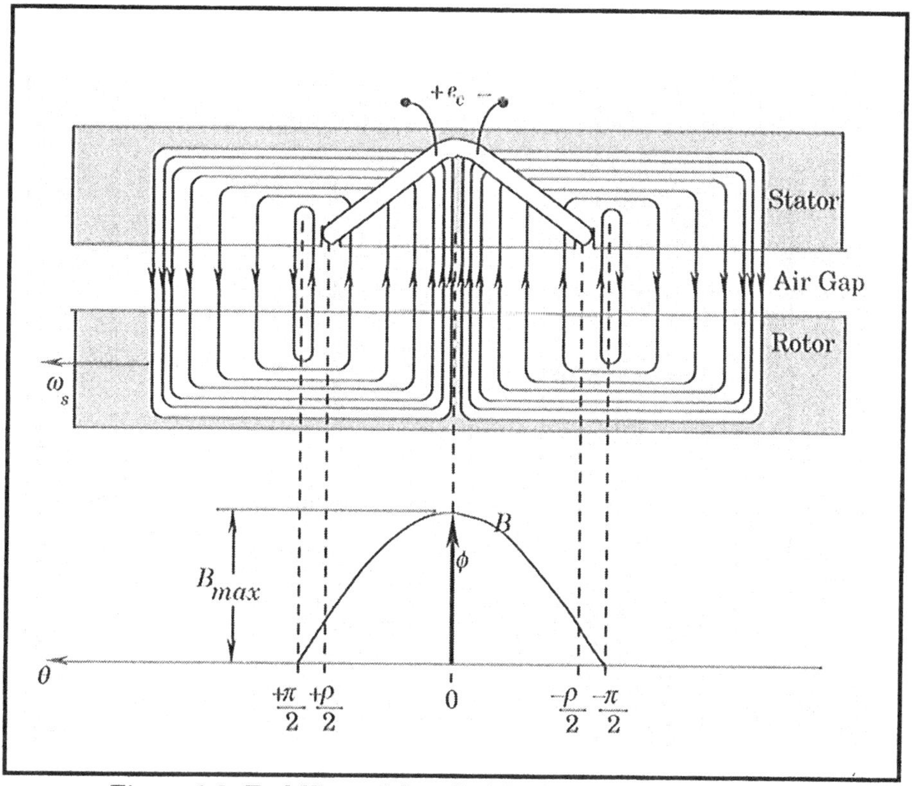

Figure 2.2 End View of One Coil Linked by Air Gap Flux

$$B = B_{\max} \cos \frac{p\theta}{2} = B_{\max} \cos \theta_e \tag{2.1}$$

where θ is the angular position in radians around the air gap in the direction indicated in Figures 2.1 and 2.2, and p is the number of poles. The angle θ_e is the same angle as θ, but measured in electrical radians. We compute the total flux linking the coil as

$$\phi_c = \iint BdA. \tag{2.2}$$

The differential area is written as

$$dA = Lrd\theta = \frac{2Lr}{p} d\theta_e \tag{2.3}$$

where L is the coil length, r is the radius of the air gap in the machine cylindrical geometry.

The generator shaft rotates at synchronous speed with velocity

$$\omega_s = \frac{2\pi f}{p/2} = \frac{2}{p}\omega_e. \tag{2.4}$$

We may write the flux density as the traveling wave

$$\begin{aligned} B(\theta,t) &= B_{\max}\cos\left[\frac{p}{2}(\theta - \omega_s t)\right] \\ &= B_{\max}\cos(\theta_e - \omega_e t). \end{aligned} \tag{2.5}$$

Substituting B and dA into the integrand and evaluating between the limits $\pm\rho/2$ we compute the total flux to be

$$\begin{aligned} \phi_c &= \int_{-\rho/2}^{+\rho/2} \frac{2Lr}{p} B_{\max} \cos(\theta_e - \omega_e t) d\theta_e \\ &= k_p \theta_p \cos \omega_e t \end{aligned} \tag{2.6}$$

where we define

$$\begin{aligned} k_p &= \text{Pitch Factor} = \sin\frac{\rho}{2} \\ \phi_p &= \text{Flux Per Pole} = \frac{4B_{\max}Lr}{p}. \end{aligned} \tag{2.7}$$

The induced electromotive force is computed from Faraday's law, which states that the emf is equal to the rate of change of flux linkages, i.e.,

$$e_c = -\frac{d\lambda_c}{dt} = -\frac{d(N_c\phi_c)}{dt} = \omega_e N_c k_p \phi_p \sin \omega_e t \tag{2.8}$$

where N_c is the number of turns in the coil. It is convenient to write the coil voltage as

$$e_c = \sqrt{2}\, E_c \sin \omega_e t \tag{2.9}$$

where E_c is the rms value of the coil voltage. Note that the total pitch of the coil $(\pi-\rho)$ is less than one pole pitch (π). This has the effect of reducing harmonics more than it reduces the fundamental component of voltage. This reduction is expressed in terms of the pitch factor. Also note that e_c is the induced voltage in only one coil, as shown in Figure 2.2.

The total voltage of one phase equals that of all coils making up the phase winding. These coils are placed in slots to form equally spaced *groups*, with the number of groups in each phase winding being equal to the number of rotor poles. The coils in the group are not all in the same slots, however, but are displaced by the slot pitch γ. Therefore the voltage induced in the individual coils will be out of phase by this angle. This means that the addition of the voltages is not a simple arithmetic addition, but is usually performed as a *phasor* addition to compute the total rms emf of the group of coils as shown in Figure 2.3, where the number of coils n in the group is assumed to be four.

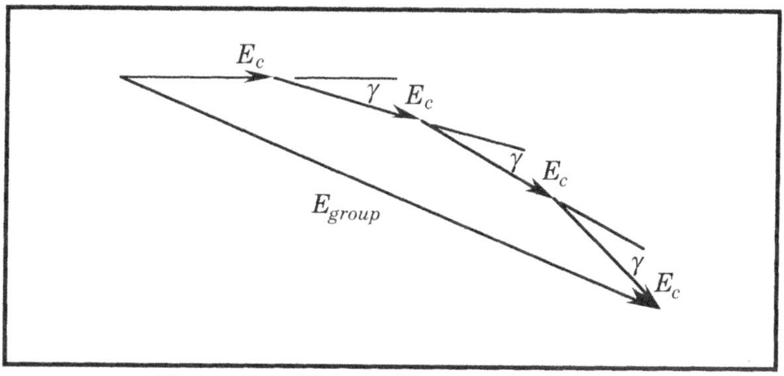

Figure 2.3 Phasor Diagram for E_{group}

From the geometry of Figure 2.3 we may compute

$$E_{group} = nE_c \frac{\sin\dfrac{n\gamma}{2}}{n\sin\dfrac{\gamma}{2}} = nE_c k_d \tag{2.10}$$

where a new constant k_d, called the distribution factor, is defined as

$$k_d = \frac{\sin\dfrac{n\gamma}{2}}{n\sin\dfrac{\gamma}{2}}. \tag{2.11}$$

Table 2.1 Defined Stator and Rotor Coils

Designation	Description of Circuit
a, b, c	stator circuits of phases a, b, and,c
F	field winding
D	d-axis amortisseur
G	q-axis field or deep amortisseur
Q	q-axis amortisseur

Finally, the phase voltage is composed of p groups in series or

$$E_{phase} = pE_{group}. \tag{2.12}$$

For steam turbine driven generators, p is usually 2 or 4. Hydro generators may have a much larger number of poles, depending on the shaft speed.

Similar equations apply for each phase and, because of the phase's 120 electrical degree displacement, gives the usual balanced three-phase induced voltages. The foregoing derivation is intended simply to justify the usual assertion that the synchronous generator produces balanced sinusoidal voltages. The interested reader should consult any elementary machinery text for a more detailed treatment of this subject [3].

We now determine the electrical properties of the stator and rotor coils so that we can derive the electric circuit behavior of the machine. In doing so, we will be primarily interested in the self and mutual inductances of the seven coils. Here, we represent the machine windings approximately as single coils. These coils are defined in Table 2.1, where rotor circuits are designated by capital letters and stator circuits by lower case letters. These letters will be used as subscripts in defining the circuit inductances.

2.2 THE MACHINE CIRCUIT INDUCTANCES

In this section we state, without proof, the self and mutual inductances of the seven circuits that make up the synchronous machine defined in Figure 2.1. A more complete development is given in [3] and [4].

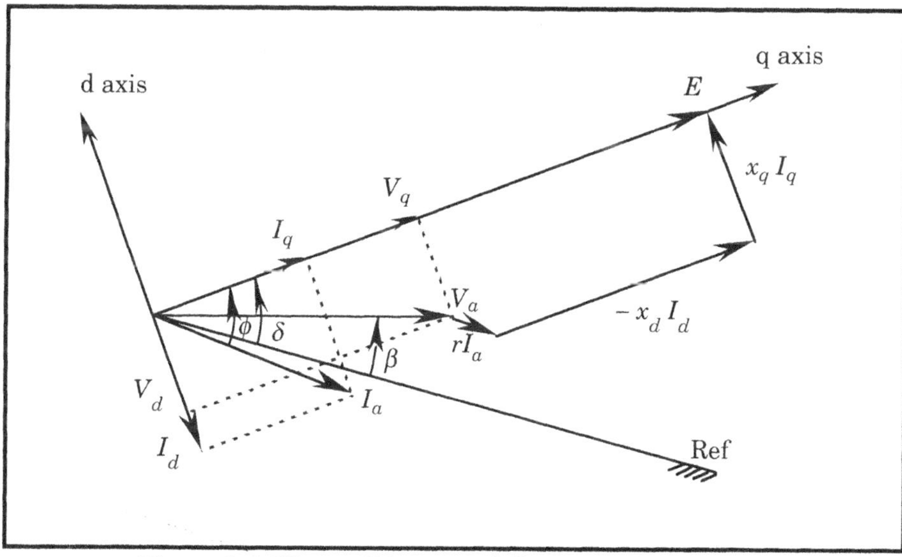

Figure 2.4 Phasor Diagram of Generator Quantities

2.2.1 Stator Self Inductances

The self inductances of the stator coils are defined as follows in mks units.

$$
\begin{aligned}
L_{aa} &= L_s + L_m \cos 2\theta && \text{H} \\
L_{bb} &= L_s + L_m \cos 2\left(\theta - \frac{2\pi}{3}\right) && \text{H} \\
L_{cc} &= L_s + L_m \cos 2\left(\theta + \frac{2\pi}{3}\right) && \text{H}
\end{aligned}
\qquad (2.13)
$$

where

$$L_s > L_m$$

and

θ = angular rotor displacement in mechanical radians

$$\theta = \omega_B t + \delta + \frac{\pi}{2}$$

and where ω_B is the base (rated) radian frequency and δ is the angle measured from a synchronously rotating reference to the q-axis. This angle and other basic quantities for the synchronous machine are shown in the phasor diagram of Figure 2.4.

See Figure 2.1 for the orientation of angular displacement. Note that both inductances on the right hand side of (2.13) are constants. The double frequency (2θ) functions occur due to the rotor saliency and the fact that the self inductances are the same for either the North or South pole of the rotor in the position shown in Figure 2.1.

2.2.2 Stator Mutual Inductances

The stator-to-stator mutual inductances are influenced by rotor saliency and therefore are a function of rotor position. From [4] we write

$$
\begin{aligned}
L_{ab} &= L_{ba} = -M_s - L_m \cos 2\left(\theta + \frac{\pi}{6}\right) \quad \text{H} \\
L_{bc} &= L_{cb} = -M_s - L_m \cos 2\left(\theta - \frac{\pi}{2}\right) \quad \text{H} \\
L_{ca} &= L_{ac} = -M_s - L_m \cos 2\left(\theta + \frac{5\pi}{6}\right) \quad \text{H}
\end{aligned}
\tag{2.14}
$$

where M_s is a constant mutual inductance. Note that the double subscript notation, with unlike subscripts, implies a mutual inductance.

2.2.3 Rotor Self Inductances

The rotor self inductances are constant. We indicate this fact (constant inductances) by simplifying the subscript notation to a single letter. In the future, this simple notation will help us to clearly identify the constant inductances in a very large number of defined quantities. Thus we write

$$
\begin{aligned}
L_{FF} &= L_F \quad \text{H} \\
L_{DD} &= L_D \quad \text{H} \\
L_{GG} &= L_G \quad \text{H} \\
L_{QQ} &= L_Q \quad \text{H.}
\end{aligned}
\tag{2.15}
$$

2.2.4 Rotor Mutual Inductances

The rotor mutual inductances are either constant or, because of their 90 degree orientation, zero. Thus we have

$$
\begin{aligned}
L_{FD} &= L_{DF} = M_X \quad \text{H} \\
L_{GQ} &= L_{QG} = M_Y \quad \text{H} \\
L_{FG} &= L_{GF} = 0 \\
L_{FQ} &= L_{QF} = 0 \\
L_{DG} &= L_{GD} = 0 \\
L_{DQ} &= L_{QD} = 0
\end{aligned}
\tag{2.16}
$$

where M_X and M_Y are positive constants.

2.2.5 Stator-to-Rotor Mutual Inductances

The stator-to-rotor mutuals may be divided into two groups – those involving the d-axis and those involving the q-axis. The mutuals involving the d-axis are given by

$$\begin{aligned} L_{aF} &= L_{Fa} = M_F \cos\theta && \text{H} \\ L_{bF} &= L_{Fb} = M_F \cos\left(\theta - \frac{2\pi}{3}\right) && \text{H} \\ L_{cF} &= L_{Fc} = M_F \cos\left(\theta + \frac{2\pi}{3}\right) && \text{H} \end{aligned} \tag{2.17}$$

$$\begin{aligned} L_{aD} &= L_{Da} = M_D \cos\theta && \text{H} \\ L_{bD} &= L_{Db} = M_D \cos\left(\theta - \frac{2\pi}{3}\right) && \text{H} \\ L_{cD} &= L_{Dc} = M_D \cos\left(\theta + \frac{2\pi}{3}\right) && \text{H} \end{aligned} \tag{2.18}$$

where M_F and M_D are positive constants.

The stator-to-rotor mutuals involving the q-axis rotor circuits are given by

$$\begin{aligned} L_{aG} &= L_{Ga} = M_G \sin\theta && \text{H} \\ L_{bG} &= L_{Gb} = M_G \sin\left(\theta - \frac{2\pi}{3}\right) && \text{H} \\ L_{cG} &= L_{Gc} = M_G \sin\left(\theta + \frac{2\pi}{3}\right) && \text{H} \end{aligned} \tag{2.19}$$

$$\begin{aligned} L_{aQ} &= L_{Qa} = M_Q \sin\theta && \text{H} \\ L_{bQ} &= L_{Qb} = M_Q \sin\left(\theta - \frac{2\pi}{3}\right) && \text{H} \\ L_{cQ} &= L_{Qc} = M_Q \sin\left(\theta + \frac{2\pi}{3}\right) && \text{H} \end{aligned} \tag{2.20}$$

where M_G and M_Q are positive constants.

This completes the specification of all self and mutual inductances for the synchronous machine.

2.3 PARK'S TRANSFORMATION

Having defined all 49 self and mutual inductances for the seven circuits, we may now write the flux linkage equation. For ease of notation, we write these equations using matrix notation. Since there are seven distinct circuits for the stator and rotor, this matrix equation will have a 7 × 7 inductance matrix, which will show clearly the coupling among all of the circuits. This matrix equation is written as follows:

$$\begin{bmatrix} \psi_a \\ \psi_b \\ \psi_c \\ \hline \psi_F \\ \psi_D \\ \hline \psi_G \\ \psi_Q \end{bmatrix} = \left[\begin{array}{ccc|cc|cc} L_{aa} & L_{ab} & L_{ac} & L_{aF} & L_{aD} & L_{aG} & L_{aQ} \\ L_{ba} & L_{bb} & L_{bc} & L_{bF} & L_{bD} & L_{bG} & L_{bQ} \\ L_{ca} & L_{cb} & L_{cc} & L_{cF} & L_{cD} & L_{cG} & L_{cQ} \\ \hline L_{Fa} & L_{Fb} & L_{Fc} & L_F & M_X & 0 & 0 \\ L_{Da} & L_{Db} & L_{Dc} & M_X & L_D & 0 & 0 \\ \hline L_{Ga} & L_{Gb} & L_{Gc} & 0 & 0 & L_G & M_Y \\ L_{Qa} & L_{Qb} & L_{Qc} & 0 & 0 & M_Y & L_Q \end{array}\right] \begin{bmatrix} i_a \\ i_b \\ i_c \\ \hline i_F \\ i_D \\ \hline i_G \\ i_Q \end{bmatrix} \tag{2.21}$$

where the units of (2.21) are Webers or Weber-turns. Note that a few inductances are constant (single subscript) and a few are zero. Most are dependent on the angular position of the rotor, as evidenced by (2.5) - (2.20), where the angular position is a function of time. Note also that (2.21) is a symmetric matrix. We simplify the notation to write (2.21) in partitioned form as

$$\begin{bmatrix} \mathbf{\Psi}_{abc} \\ \hline \mathbf{\Psi}_{FD} \\ \hline \mathbf{\Psi}_{GQ} \end{bmatrix} = \left[\begin{array}{c|c|c} \mathbf{L}_{SS} & \mathbf{L}_{SD} & \mathbf{L}_{SQ} \\ \hline \mathbf{L}_{SD}^{\mathrm{T}} & \mathbf{L}_{DD} & \mathbf{0} \\ \hline \mathbf{L}_{SQ}^{\mathrm{T}} & \mathbf{0} & \mathbf{L}_{QQ} \end{array}\right] \begin{bmatrix} \mathbf{i}_{abc} \\ \hline \mathbf{i}_{FD} \\ \hline \mathbf{i}_{GQ} \end{bmatrix}. \tag{2.22}$$

Note that this matrix has a nearly diagonal form and that the lower right portion (DD and QQ) contains only constant matrices (see the single subscripts in equation 2.21). The matrix in the SS position is dependent on angular position, θ, and time. We seek a means of simplifying this matrix, particularly the time-varying partition in the upper left corner. The desired simplification is accomplished by means of a transformation of variables from the a-b-c frame of reference to a new reference frame. We call this transformation "Park's transformation," after R. H. Park [6,7].

The procedure for diagonalizing a matrix is well known [8]. Indeed, if **A** is a real $n \times n$ symmetric matrix, there exists an orthogonal $n \times n$ transformation matrix **Q** such that

$$\mathbf{D} = \mathbf{Q}^t \mathbf{A} \mathbf{Q} \tag{2.23}$$

is a diagonal matrix **D** whose elements are the eigenvalues of the matrix **A**. The procedure requires, first, the calculation of the eigenvalues of **A**. From these eigenvalues, we compute the eigenvectors. If the eigenvectors are distinct, these eigenvectors form an orthogonal basis for the new reference frame and become the columns of the desired transformation matrix **Q**.

For the synchronous machine stator inductance matrix we compute the eigenvalues by a straightforward procedure [8]. First, we write

$$\det\left[\lambda \mathbf{U}_3 - \mathbf{L}_{SS}\right] = 0 \tag{2.24}$$

where

$\mathbf{L}_{SS}$ = the stator inductance matrix from (2.22)
$\mathbf{U}_3$ = a 3×3 unit or identity matrix
λ = the eigenvalue variable.

Since the stator inductance matrix is 3×3, equation (2.24) is a cubic equation in λ, given by

$$\begin{aligned} &\lambda^3 - \left(L_{aa} + L_{bb} + L_{cc}\right)\lambda^2 \\ &+ \left(L_{aa}L_{bb} + L_{bb}L_{cc} + L_{cc}L_{aa} - L_{ab}^2 - L_{bc}^2 - L_{ca}^2\right)\lambda \\ &+ \left(L_{aa}L_{bc}^2 + L_{bb}L_{ca}^2 + L_{cc}L_{ab}^2 - L_{aa}L_{bb}L_{cc} - 2L_{ab}L_{bc}L_{ca}\right) = 0 \end{aligned} \tag{2.25}$$

or, in factored form

$$\left(\lambda - \lambda_1\right)\left(\lambda - \lambda_2\right)\left(\lambda - \lambda_3\right) = 0 \tag{2.26}$$

where $\lambda_1, \lambda_2, \lambda_3$ are the three eigenvalues.

Equation (2.25) can be factored in the form of (2.26), using the inductance definitions of (2.5) to (2.20). This laborious task gives the simple result:

$$\lambda_1 = L_s - 2M_s$$
$$\lambda_2 = L_s + M_s + \frac{3}{2}L_m$$
$$\lambda_3 = L_s + M_s - \frac{3}{2}L_m. \tag{2.27}$$

Note that the eigenvalues are constant (single subscript) and are not functions of either time or angular rotor position.

To compute the eigenvectors, we solve the equation

$$(\mathbf{L}_{SS} - \lambda_i \mathbf{U}_3)\mathbf{v}_i = 0, \qquad i = 1,2,3 \tag{2.28}$$

where $\mathbf{v}_i$ is the eigenvector corresponding to λ_i.

For the first eigenvalue we find the eigenvector

$$\bar{\mathbf{v}}_1 = \begin{bmatrix} v \\ v \\ v \end{bmatrix} \tag{2.29}$$

which has

$$Length = v\sqrt{3}.$$

We normalize the eigenvector by dividing all its elements by the length to compute the normalized eigenvector

$$\mathbf{v}_1 = \begin{bmatrix} 1/\sqrt{3} \\ 1/\sqrt{3} \\ 1/\sqrt{3} \end{bmatrix}. \tag{2.30}$$

For the second eigenvalue, we again apply (2.28) and normalize the result to compute

$$\mathbf{v}_2 = \sqrt{\frac{2}{3}}\begin{bmatrix} \cos\theta \\ \cos(\theta - 2\pi/3) \\ \cos(\theta + 2\pi/3) \end{bmatrix}. \tag{2.31}$$

Finally, for the third eigenvalue we compute the normalized result

$$\mathbf{v}_3 = \sqrt{\frac{2}{3}}\begin{bmatrix} \sin\theta \\ \sin(\theta - 2\pi/3) \\ \sin(\theta + 2\pi/3) \end{bmatrix}. \tag{2.32}$$

Then we may compute $\mathbf{Q}$ as

$$\mathbf{Q} = \begin{bmatrix} \mathbf{v}_1 & \mathbf{v}_2 & \mathbf{v}_3 \end{bmatrix} \tag{2.33}$$

and we may easily show that

$$\mathbf{Q}^{\mathrm{T}} = \mathbf{Q}^{-1}. \tag{2.34}$$

Also, we can verify, by straightforward algebraic manipulation, that

$$\mathbf{Q}^{\mathrm{T}}\mathbf{L}_{ss}\mathbf{Q} = \mathrm{diag}(\lambda_1, \lambda_2, \lambda_3) \tag{2.35}$$

exactly as the theory prescribes.

In the notation of synchronous machine theory, we give these eigenvalues unique designations, namely,

$$\begin{aligned} L_0 &= \lambda_1 = L_s - 2M_s \\ L_d &= \lambda_2 = L_s + M_s + \frac{3}{2}L_m \\ L_q &= \lambda_3 = L_s + M_s - \frac{3}{2}L_m. \end{aligned} \tag{2.36}$$

We now define a transformation matrix that we shall call the Park's transformation $\mathbf{P}$, which is given by

$$\mathbf{P} = \mathbf{Q}^{-1} = \begin{bmatrix} \frac{1}{\sqrt{3}} & \frac{1}{\sqrt{3}} & \frac{1}{\sqrt{3}} \\ \sqrt{\frac{2}{3}}\cos\theta & \sqrt{\frac{2}{3}}\cos\left(\theta - \frac{2\pi}{3}\right) & \sqrt{\frac{2}{3}}\cos\left(\theta + \frac{2\pi}{3}\right) \\ \sqrt{\frac{2}{3}}\sin\theta & \sqrt{\frac{2}{3}}\sin\left(\theta - \frac{2\pi}{3}\right) & \sqrt{\frac{2}{3}}\sin\left(\theta + \frac{2\pi}{3}\right) \end{bmatrix} \tag{2.37}$$

such that

$$\mathbf{PL}_{SS}\mathbf{P}^{\mathrm{T}} = \mathrm{diag}(L_0, L_d, L_q) \tag{2.38}$$

where, since $\mathbf{P}$ is orthogonal, we note that

$$\mathbf{P}^{\mathrm{T}} = \mathbf{P}^{-1}. \tag{2.39}$$

Now, from (2.22), we have a 7×7 matrix equation. We shall premultiply both sides of (2.22) by the 7×7 transformation matrix

$$\mathbf{T} = \begin{bmatrix} \mathbf{P} & \mathbf{0} & \mathbf{0} \\ \mathbf{0} & \mathbf{U}_2 & \mathbf{0} \\ \mathbf{0} & \mathbf{0} & \mathbf{U}_2 \end{bmatrix} \tag{2.40}$$

where

$\mathbf{U}_2$ = a 2×2 unit matrix
$\mathbf{P}$ = the Park's transformation matrix.

The result is given by

$$\begin{bmatrix} \boldsymbol{\Psi}_{0dq} \\ \hline \boldsymbol{\Psi}_{FD} \\ \hline \boldsymbol{\Psi}_{GQ} \end{bmatrix} = \left[\begin{array}{c|c|c} \mathbf{PL}_{SS}\mathbf{P}^{-1} & \mathbf{PL}_{SD} & \mathbf{PL}_{SQ} \\ \hline \mathbf{L}_{SD}^{\mathrm{T}}\mathbf{P}^{-1} & \mathbf{L}_{DD} & \mathbf{0} \\ \hline \mathbf{L}_{SQ}^{\mathrm{T}}\mathbf{P}^{-1} & \mathbf{0} & \mathbf{L}_{QQ} \end{array}\right] \begin{bmatrix} \mathbf{i}_{0dq} \\ \hline \mathbf{i}_{FD} \\ \hline \mathbf{i}_{GQ} \end{bmatrix} \tag{2.41}$$

where, by definition,

$$\mathbf{P}\boldsymbol{\psi}_{abc} = \mathbf{P}\begin{bmatrix} \psi_a \\ \psi_b \\ \psi_c \end{bmatrix} = \begin{bmatrix} \psi_0 \\ \psi_d \\ \psi_q \end{bmatrix} = \boldsymbol{\Psi}_{0dq} \tag{2.42}$$

and similarly for currents and voltages.

Now, consider the transformed inductance matrices in (2.41). We have already determined the upper left term of (2.41), with the result given by (2.38). We may also readily verify that

$$\mathbf{PL}_{SD} = \begin{bmatrix} 0 & 0 \\ kM_F & kM_D \\ 0 & 0 \end{bmatrix} \tag{2.43}$$

where

$$k = \sqrt{\frac{3}{2}}. \tag{2.44}$$

Also we may compute

$$\mathbf{PL}_{SQ} = \begin{bmatrix} 0 & 0 \\ 0 & 0 \\ kM_G & kM_Q \end{bmatrix}. \tag{2.45}$$

Finally we note that

$$\begin{aligned} \mathbf{L}_{SD}^{T}\mathbf{P}^{-1} &= \left(\mathbf{PL}_{SQ}\right)^{T} \\ \mathbf{L}_{SQ}^{T}\mathbf{P}^{-1} &= \left(\mathbf{PL}_{SD}\right)^{T} \end{aligned} \tag{2.46}$$

so that these partitions of (2.41) are determined by taking the transpose of (2.43) and (2.45).

The completed transformation is given by

$$\begin{bmatrix} \psi_0 \\ \psi_d \\ \psi_q \\ \hline \psi_F \\ \psi_D \\ \hline \psi_G \\ \psi_Q \end{bmatrix} = \left[\begin{array}{ccc|cc|cc} L_0 & & & & & & \\ & L_d & & kM_F & kM_D & & \\ & & L_q & & & kM_G & kM_Q \\ \hline & kM_F & & L_F & M_X & & \\ & kM_D & & M_X & L_D & & \\ \hline & & kM_G & & & L_G & M_Y \\ & & kM_Q & & & M_Y & L_Q \end{array}\right] \begin{bmatrix} i_0 \\ i_d \\ i_q \\ \hline i_F \\ i_D \\ \hline i_G \\ i_Q \end{bmatrix} \tag{2.47}$$

where the zero items have been left blank to emphasize the sparsity of the matrix.

We note the following concerning (2.47):

1. The new inductance matrix is symmetric;

2. The matrix is constant--note that all inductances have a single subscript ;

3. ψ_0 is completely decoupled from ψ_d and ψ_q, i.e.,

   $\psi_0 = L_0 i_0$ is not a function of any current except i_0 ;

4. The units of (2.47) are Weber-turns, with inductances in henrys and currents in amperes;

5. The constant k, defined by (2.44), comes from the way in which the Park transformation was defined, and from the requirement that the transformation be power invariant.

It is helpful to rearrange the flux linkage equation to the following form

$$\begin{bmatrix} \psi_0 \\ \hline \psi_d \\ \psi_F \\ \psi_D \\ \hline \psi_q \\ \psi_G \\ \psi_Q \end{bmatrix} = \left[\begin{array}{c|ccc|ccc} L_0 & & & & & & \\ \hline & L_d & kM_F & kM_D & & & \\ & kM_F & L_F & M_X & & & \\ & kM_D & M_X & L_D & & & \\ \hline & & & & L_q & kM_G & kM_Q \\ & & & & kM_G & L_G & M_Y \\ & & & & kM_Q & M_Y & L_Q \end{array}\right] \begin{bmatrix} i_0 \\ \hline i_d \\ i_F \\ i_D \\ \hline i_q \\ i_G \\ i_Q \end{bmatrix}. \tag{2.48}$$

This rearrangement shows more clearly the decoupling of the three circuits. We may now easily derive equivalent circuits for the equations of (2.48). This circuit is given in Figure 2.5. Note that the self and mutual inductances are all constants and are not dependent on rotor position. Since the rotor circuits are unaffected by the transformation, we conclude that the new d and q circuits are stator equivalent circuits that move as if attached to the rotor and with physical orientation aligned exactly with the d- and q-axes. The circuit subscripted with the zero (0) has no mutual coupling with either the d- or q-axis circuits and is therefore in quadrature with the d- and q-axes. This third circuit must be orthogonal to the d- and q-axes. It therefore magnetizes an axis that lies along the rotor centerline or rotational axis and is perpendicular to the plane formed by the d- and q-

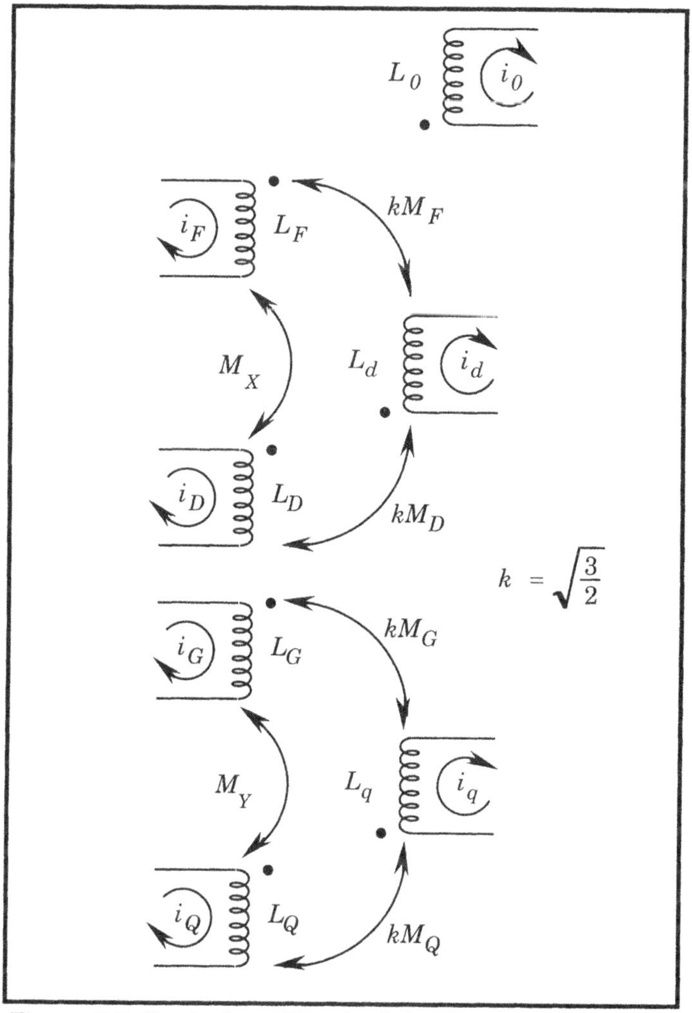

Figure 2.5 Equivalent Circuit of the Transformed Stator and Rotor Coupled Circuits

axes. We shall see later that this third circuit is exactly the zero sequence circuit and has zero current under balanced loading conditions.

2.4 THE VOLTAGE EQUATIONS

The voltage equations of the synchronous generator are written in reference to Figure 2.6. By direct application of Kirchhoff's laws we write

$$\begin{bmatrix} v_a \\ v_b \\ v_c \\ \hline -v_F \\ -v_D \\ \hline -v_G \\ -v_Q \end{bmatrix} = \left[\begin{array}{ccc|cc|cc} r_a & & & & & & \\ & r_b & & & & & \\ & & r_c & & & & \\ \hline & & & r_F & & & \\ & & & & r_D & & \\ \hline & & & & & r_G & \\ & & & & & & r_Q \end{array}\right] \begin{bmatrix} i_a \\ i_b \\ i_c \\ \hline i_F \\ i_D \\ \hline i_G \\ i_Q \end{bmatrix} - \begin{bmatrix} p\psi_a \\ p\psi_b \\ p\psi_c \\ \hline p\psi_F \\ p\psi_D \\ \hline p\psi_G \\ p\psi_Q \end{bmatrix} + \begin{bmatrix} v_n \\ v_n \\ v_n \\ \hline 0 \\ 0 \\ \hline 0 \\ 0 \end{bmatrix} \text{ V} \tag{2.49}$$

where we use the operator $p = d/dt$. This equation can be written in matrix form with clear partitions for stator and rotor, as follows.

$$\begin{bmatrix} \mathbf{v}_{abc} \\ \mathbf{v}_R \end{bmatrix} = -\begin{bmatrix} \mathbf{R}_S & \mathbf{0} \\ \mathbf{0} & \mathbf{R}_R \end{bmatrix} \begin{bmatrix} \mathbf{i}_{abc} \\ \mathbf{i}_R \end{bmatrix} - \begin{bmatrix} p\boldsymbol{\psi}_{abc} \\ p\boldsymbol{\psi}_R \end{bmatrix} + \begin{bmatrix} \mathbf{v}_n \\ \mathbf{0} \end{bmatrix} \tag{2.50}$$

where we use the subscript "R" to designate all rotor circuits and either "abc" or "S" to designate all stator circuits.

We may transform the stator partition of (2.50) from the abc frame of reference to $0dq$ by premultiplying (2.50) by the transformation matrix $\mathbf{T}$, which we write as

$$\mathbf{T} = \begin{bmatrix} \mathbf{P} & \mathbf{0} \\ \mathbf{0} & \mathbf{U} \end{bmatrix} \tag{2.51}$$

where $\mathbf{P}$ is the Park's transformation matrix (2.37) and $\mathbf{U}$ is a 4×4 unit matrix. Thus we compute

$$\begin{bmatrix} \mathbf{P} & \mathbf{0} \\ \mathbf{0} & \mathbf{U} \end{bmatrix} \begin{bmatrix} \mathbf{v}_{abc} \\ \mathbf{v}_R \end{bmatrix} = -\begin{bmatrix} \mathbf{P} & \mathbf{0} \\ \mathbf{0} & \mathbf{U} \end{bmatrix} \begin{bmatrix} \mathbf{R}_S & \mathbf{0} \\ \mathbf{0} & \mathbf{R}_R \end{bmatrix} \begin{bmatrix} \mathbf{P}^{-1} & \mathbf{0} \\ \mathbf{0} & \mathbf{U} \end{bmatrix} \begin{bmatrix} \mathbf{P} & \mathbf{0} \\ \mathbf{0} & \mathbf{U} \end{bmatrix} \begin{bmatrix} \mathbf{i}_{abc} \\ \mathbf{i}_R \end{bmatrix}$$

$$-\begin{bmatrix} \mathbf{P} & \mathbf{0} \\ \mathbf{0} & \mathbf{U} \end{bmatrix} \begin{bmatrix} p\boldsymbol{\psi}_{abc} \\ p\boldsymbol{\psi}_R \end{bmatrix} + \begin{bmatrix} \mathbf{P} & \mathbf{0} \\ \mathbf{0} & \mathbf{U} \end{bmatrix} \begin{bmatrix} \mathbf{v}_n \\ \mathbf{0} \end{bmatrix}. \tag{2.52}$$

Note that we insert the product of transformation (2.51) and its inverse following the resistance matrix. This product is the identity matrix and makes not change in the equation.

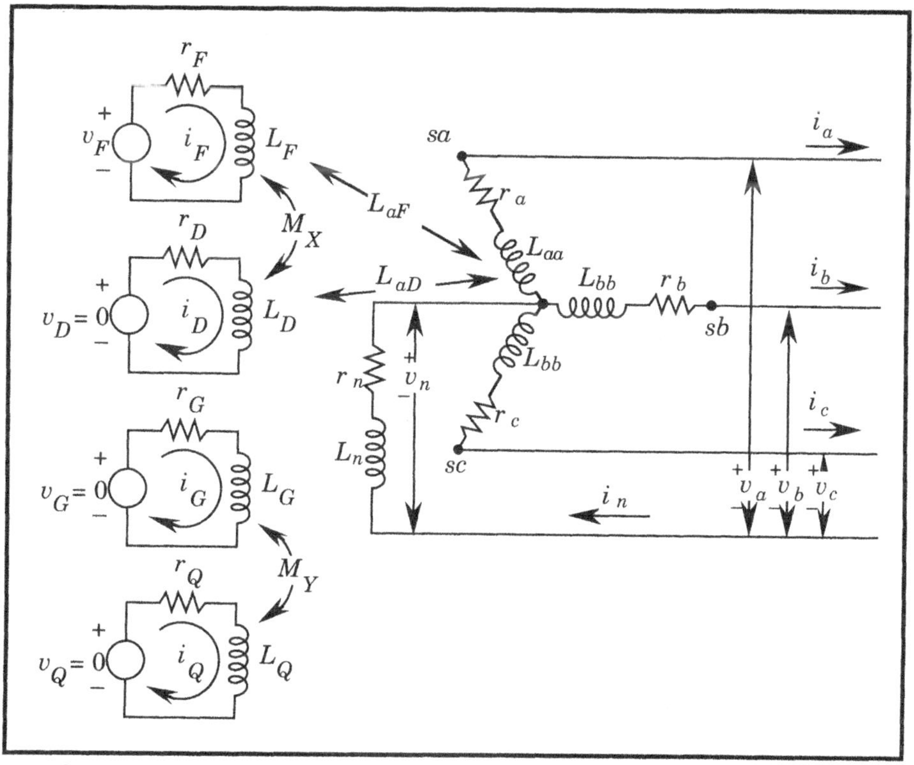

Figure 2.6 Circuit Representation of the Synchronous Generator in the a - b - c Frame of Reference

Carrying out the indicated matrix operations, we have

$$\begin{bmatrix} \mathbf{v}_{0dq} \\ \mathbf{v}_R \end{bmatrix} = -\begin{bmatrix} \mathbf{PR}_S\mathbf{P}^{-1} & \mathbf{0} \\ \mathbf{0} & \mathbf{R}_R \end{bmatrix}\begin{bmatrix} \mathbf{i}_{0dq} \\ \mathbf{i}_R \end{bmatrix} - \begin{bmatrix} \mathbf{P}p\boldsymbol{\psi}_{abc} \\ p\boldsymbol{\psi}_R \end{bmatrix} + \begin{bmatrix} \mathbf{Pv}_n \\ \mathbf{0} \end{bmatrix}. \tag{2.35}$$

We now evaluate the submatrices that are functions of $\mathbf{P}$. We can easily show that, for the practical case where

$$r_a = r_b = r_c \tag{2.54}$$

then

$$\mathbf{PR}_S\mathbf{P}^{-1} = \mathbf{R}_S. \tag{2.55}$$

Also, we may compute

$$\mathbf{Pv}_n = -\begin{bmatrix} 3r_n i_0 \\ 0 \\ 0 \end{bmatrix} - \begin{bmatrix} 3L_n\, p i_0 \\ 0 \\ 0 \end{bmatrix} = \mathbf{v}_{n0} \tag{2.56}$$

where we give this result a new variable name for convenience.

The term $\mathbf{P}(p\boldsymbol{\psi}_{abc})$ in (2.53) requires more detailed examination. From the definition of the Park's transformation

$$\boldsymbol{\psi}_{0dq} = \mathbf{P}\boldsymbol{\psi}_{abc} \tag{2.57}$$

we compute the derivative with respect to time of (2.57) as

$$p\boldsymbol{\psi}_{0dq} = \mathbf{P}(p\boldsymbol{\psi}_{abc}) + (p\mathbf{P})\boldsymbol{\psi}_{abc} \tag{2.58}$$

or, rearranging the terms,

$$\begin{aligned} \mathbf{P}(p\boldsymbol{\psi}_{abc}) &= p\boldsymbol{\psi}_{0dq} - (p\mathbf{P})\boldsymbol{\psi}_{abc} \\ &= p\boldsymbol{\psi}_{0dq} - (p\mathbf{P})\mathbf{P}^{-1}\boldsymbol{\psi}_{0dq}. \end{aligned} \tag{2.59}$$

Now, we can easily show that

$$(p\mathbf{P})\mathbf{P}^{-1} = \begin{bmatrix} 0 & 0 & 0 \\ 0 & 0 & -\omega \\ 0 & +\omega & 0 \end{bmatrix}. \tag{2.60}$$

Then we define the speed voltage vector $\mathbf{v}_\omega$ as follows

$$\mathbf{v}_\omega = (p\mathbf{P})\mathbf{P}^{-1}\boldsymbol{\psi}_{0dq} = \begin{bmatrix} 0 \\ -\omega\psi_q \\ +\omega\psi_d \end{bmatrix}. \tag{2.61}$$

Note that there is no speed voltage in the zero sequence network. Finally then, (2.59) may be written as

$$\mathbf{P}(p\boldsymbol{\psi}_{abc}) = p\boldsymbol{\psi}_{0dq} - \mathbf{v}_\omega \tag{2.62}$$

and (2.53) becomes

$$\begin{bmatrix} \mathbf{v}_{0dq} \\ \mathbf{v}_R \end{bmatrix} = -\begin{bmatrix} \mathbf{R}_S & \mathbf{0} \\ \mathbf{0} & \mathbf{R}_R \end{bmatrix}\begin{bmatrix} \mathbf{i}_{0dq} \\ \mathbf{i}_R \end{bmatrix} - \begin{bmatrix} p\boldsymbol{\Psi}_{0dq} \\ p\boldsymbol{\Psi}_R \end{bmatrix} + \begin{bmatrix} \mathbf{v}_{n0} \\ \mathbf{0} \end{bmatrix} + \begin{bmatrix} \mathbf{v}_{\omega} \\ 0 \end{bmatrix}. \tag{2.63}$$

rearranging the equations so that all equations of a given circuit are grouped together we may write

$$\begin{bmatrix} +v_0 \\ \hline +v_d \\ -v_F \\ -v_D \\ \hline +v_q \\ -v_G \\ -v_Q \end{bmatrix} = -\left[\begin{array}{c|ccc|ccc} r_0+3r_n & & & & & & \\ \hline & r_a & & & & & \\ & & r_F & & & & \\ & & & r_D & & & \\ \hline & & & & r_a & & \\ & & & & & r_G & \\ & & & & & & r_Q \end{array}\right]\begin{bmatrix} i_0 \\ \hline i_d \\ i_F \\ i_D \\ \hline i_q \\ i_G \\ i_Q \end{bmatrix}$$

$$-\left[\begin{array}{c|ccc|ccc} L_0+3L_n & & & & & & \\ \hline & L_d & kM_F & kM_D & & & \\ & kM_F & L_F & M_X & & & \\ & kM_D & M_X & L_D & & & \\ \hline & & & & L_q & kM_G & kM_Q \\ & & & & kM_G & L_G & M_Y \\ & & & & kM_Q & M_Y & L_Q \end{array}\right]\begin{bmatrix} pi_0 \\ \hline pi_d \\ pi_F \\ pi_D \\ \hline pi_q \\ pi_G \\ pi_Q \end{bmatrix} + \begin{bmatrix} 0 \\ \hline -\omega\psi_q \\ 0 \\ 0 \\ \hline +\omega\psi_d \\ 0 \\ 0 \end{bmatrix} \tag{2.64}$$

where all quantities are in mks units and $p = d/dt$ with t in seconds.

In writing (2.64), we have made use of (2.48) to write the speed voltage terms as

$$\omega\psi_d = \omega\left(L_d i_d + kM_F i_F + kM_D i_D\right)$$

$$\omega\psi_q = \omega\left(L_q i_q + kM_G i_G + kM_Q i_Q\right). \tag{2.65}$$

Equation (2.64) may be represented by the $0dq$ equivalent circuit shown in Figure 2.7, where we also note that the damper winding driving voltages are zero (these voltages are carried symbolically in (2.64) for the sake of completeness of notation).

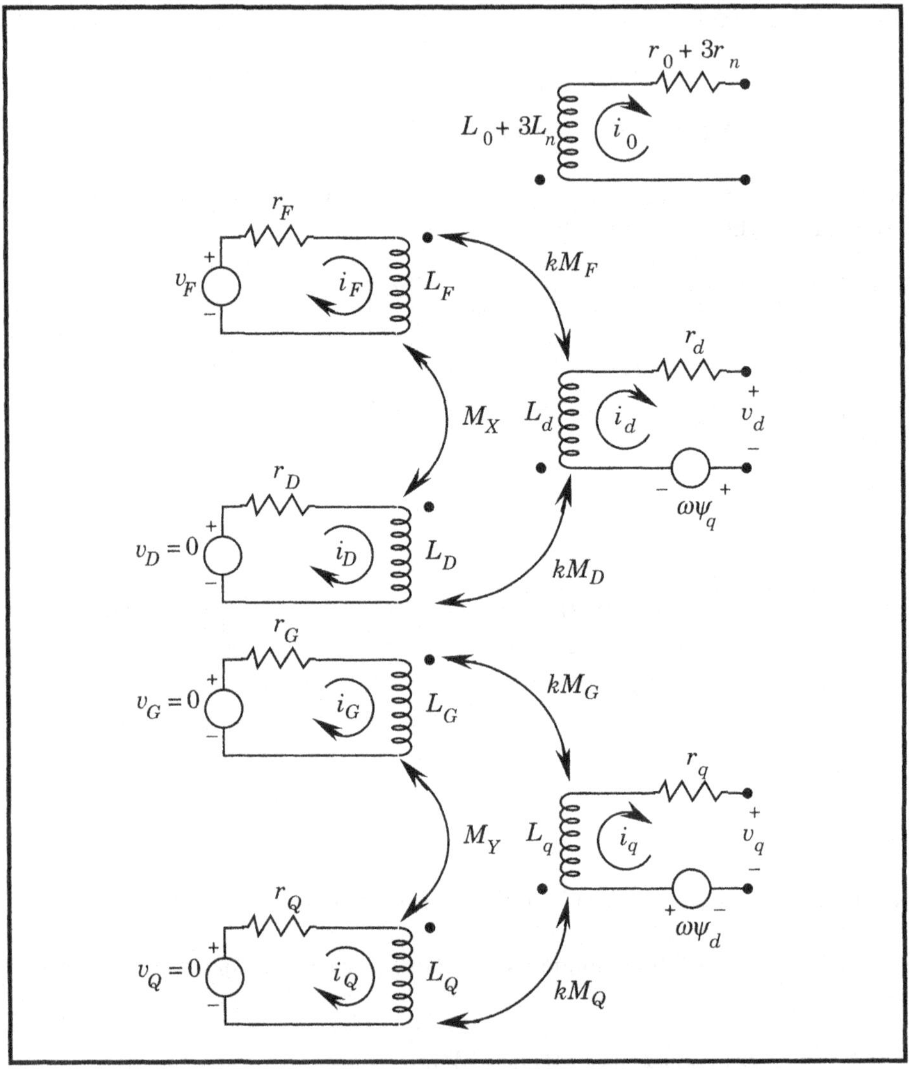

Figure 2.7 Circuit Representation of the Synchronous Machine in the 0-d-q Frame of Reference

This circuit is much simpler than that of Figure 2.6. Note that all inductances are constant. Moreover, the zero sequence network is completely decoupled and can be neglected when studying balanced conditions. The price that we pay for this simplification is the introduction of speed voltage equations, which appear in the circuit diagram as controlled sources (or more precisely as "current-controlled voltage sources"). This is important. The d and q circuits are not really decoupled because of the speed voltage terms, represented by these controlled sources. The d-axis speed voltage depends on the q-axis currents, and vice versa. These speed voltages also depend on the speed of the shaft, ω, which is not a constant under transient conditions. Hence, the speed voltage terms are nonlinear.

The rotor applied voltages are usually all zero except for the field voltage, which is due to the excitation system. A few machines are doubly excited, with dc sources applied at both the F and G windings. These machines can be analyzed using the same equations as given above if one introduces the second source of excitation to the G winding.

2.5 THE POWER AND TORQUE EQUATIONS

To develop the power and torque equations for the synchronous generator, we begin with a basic energy balance concept.

1. mechanical energy
    - energy transferred mechanically
    - energy loss through friction and windage.
2. electrical energy
    - energy transferred through circuits
    - energy stored in the fields of inductances
    - energy ohmic loss.
3. field energy
    - energy transferred through the field
    - energy stored in the magnetic field
    - energy loss due to hysteresis and eddy currents.

Thus, we write the general energy balance equation as

$$\begin{bmatrix}\text{Mechanical}\\ \text{Energy}\\ \text{Input}\end{bmatrix} - \begin{bmatrix}\text{Friction \&}\\ \text{Windage}\\ \text{Energy Loss}\end{bmatrix} = \begin{bmatrix}\text{Increase in}\\ \text{Field Stored}\\ \text{Energy}\end{bmatrix} - \begin{bmatrix}\text{Field}\\ \text{Heat}\\ \text{Loss}\end{bmatrix} + \begin{bmatrix}\text{Electrical}\\ \text{Energy}\\ \text{Output}\end{bmatrix} - \begin{bmatrix}\text{Electrical}\\ \text{Ohmic}\\ \text{Loss}\end{bmatrix}. \tag{2.66}$$

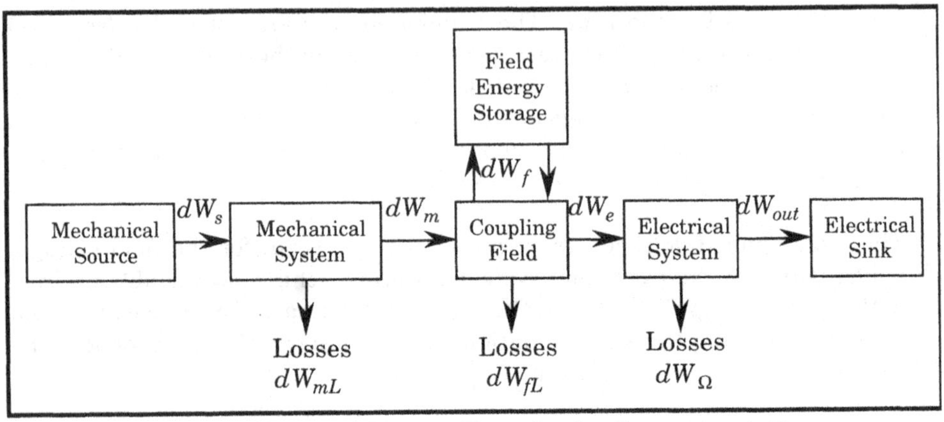

Figure 2.8 Differential Energy Transfer for Generator Action

It is convenient to diagram this process as shown in Figure 2.8.

By inspection of Figure 2.8 we write

$$dW_m = dW_s - dW_{mL} = dW_f + dW_{fL} + dW_e. \tag{2.67}$$

The differential energy terms associated with the field losses are the hysteresis and eddy current losses that are common in ferromagnetic materials. These losses are usually associated with the mechanical losses to create a total loss term due to mechanical and field effects called the "rotational losses,"

$$dW_{rot} = dW_{mL} + dW_{fL}. \tag{2.68}$$

This artificial grouping of losses is justified since the field losses are small and play no role in the basic energy conversion process. Hence, the field losses are divorced from the field stored energy term. The result of this grouping gives the "internal differential mechanical energy," which is given by

$$dW_m = dW_s - dW_{rot} = dW_f + dW_e \tag{2.69}$$

where we recognize that this mechanical energy is the "net" energy available for conversion. Each increment of mechanical energy injected

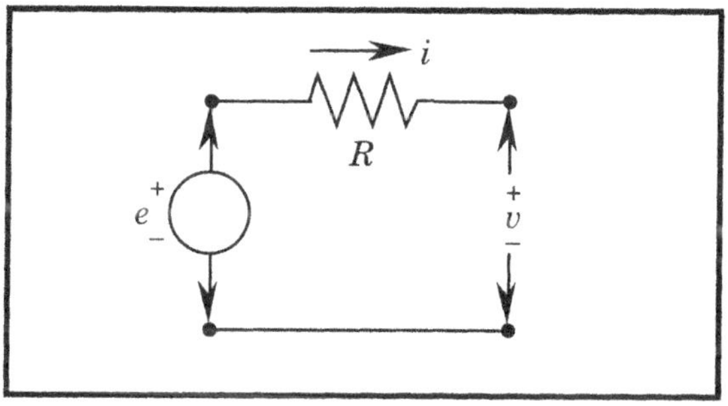

Figure 2.9 Equivalent Circuit of the Generator Terminal

may be converted into stored energy in the magnetic field or into electrical energy. We also recognize that, in the steady state, the synchronous machine is a constant field machine, hence the differential field energy is zero. Energy is exchanged with the field storage medium during transients, after which the differential field energy term again goes to zero.

The electrical system depicted in Figure 2.8 is only that part of the system that separates the losses from the electrical output. It is convenient to think of this in terms of an equivalent circuit as shown in Figure 2.9, where we recognize the presence of an internal induced emf e, which is created by reaction with the magnetic field as a $B\ell v$ induction. By inspection, then

$$\begin{aligned} dW_e &= ei\,dt = (v + Ri)i\,dt \\ &= vi\,dt + Ri^2\,dt \\ &= dW_{out} + dW_\Omega. \end{aligned} \tag{2.70}$$

Thus, the basic equation (2.69) may be written as

$$dW_{out} = dW_m - dW_f - dW_\Omega. \tag{2.71}$$

The time derivative of (2.71) is the power equation

$$\frac{dW_{out}}{dt} = \frac{dW_m}{dt} - \frac{dW_f}{dt} - \frac{dW_\Omega}{dt} \tag{2.72}$$

or

$$p_{out} = p_m - p_f - p_\Omega. \tag{2.73}$$

The instantaneous power output of a three-phase synchronous generator is given by the equation

$$
\begin{aligned}
p_{out} &= v_a i_a + v_b i_b + v_c i_c \\
&= \mathbf{v}_{abc}^{\mathrm{T}} \mathbf{i}_{abc}
\end{aligned}
\tag{2.74}
$$

but

$$
\begin{aligned}
\mathbf{i}_{abc} &= \mathbf{P}^{-1} \mathbf{i}_{0dq} \\
\mathbf{v}_{abc} &= \mathbf{P}^{-1} \mathbf{v}_{0dq}.
\end{aligned}
\tag{2.75}
$$

Substituting (2.75) into (2.74), we compute

$$
\begin{aligned}
p_{out} &= \mathbf{v}_{0dq}^{\mathrm{T}} \left(\mathbf{P}^{-1}\right)^{\mathrm{T}} \mathbf{P}^{-1} \mathbf{i}_{0dq} \\
&= \mathbf{v}_{0dq}^{\mathrm{T}} \mathbf{i}_{0dq}.
\end{aligned}
\tag{2.76}
$$

Thus the power is invariant under the Park's transformation as defined in this monograph. This could have been predicted since **P** is orthogonal. (Note: the **P** transformation was chosen to be orthogonal. This was not true of the transformation originally used by R. H. Park, which resulted in (2.76) being multiplied by a constant and also resulted in a nonsymmetric inductance matrix).

Substituting for the voltages from (2.64), we write the output power expression as

$$
\begin{aligned}
p_{out} = &- r_a \left(i_0^2 + i_d^2 + i_q^2 \right) - 3 r_n i_0^2 \\
&- \left(i_0 p\psi_0 + i_d p\psi_d + i_q p\psi_q \right) \\
&+ \omega \left(\psi_d i_q - \psi_q i_d \right) \quad \mathrm{W}
\end{aligned}
\tag{2.77}
$$

or

$$
p_{out} = - p_\Omega - p_f + p_m \tag{2.78}
$$

where

p_Ω = ohmic losses

p_f = rate of increased energy storage in the fields

p_m = mechanical power transferred across the air gap.

It is convenient to define the electrical power

$$
p_e = p_m - p_f \quad \mathrm{W} \tag{2.79}
$$

which is the net power transmitted to the electrical system of the machine. Then

$$p_{out} = p_e - p_\Omega \qquad \text{W} \tag{2.80}$$

and we may think of the electrical power as an internal source of power, behind the stator resistance. In practice, we often set the electrical and air gap (mechanical) powers equal since the rate of change of field energy is very small compared to the air gap power.

Torque is obtained by dividing power by angular velocity. Thus we have synchronous machine:torque equation

$$T_m = \psi_d i_q - \psi_q i_d \qquad \text{N m} \tag{2.81}$$

and, since

$$\omega = \frac{d\theta}{dt} \tag{2.82}$$

we write the torque corresponding to the field energy transfer as

$$T_f = i_0 \frac{d\psi_0}{d\theta} + i_d \frac{d\psi_d}{d\theta} + i_q \frac{d\psi_q}{d\theta} \tag{2.83}$$

and

$$T_e = T_m - T_f \quad \text{Nm.} \tag{2.84}$$

Again, we note that the last term is usually very small and is often neglected.

2.6 NORMALIZATION OF THE EQUATIONS

The voltage equations that describe the synchronous generator model, given by (2.64), are all in mks units with voltages in volts, currents in amperes, resistances in ohms, inductances in henrys, and flux linkages in webers. It is common, however, for these equations to be normalized and expressed in per unit, based on some arbitrarily chosen, but coordinated, base quantities.

In normalizing any system of equations, it is necessary to observe the dimensional constraints among the quantities. These constraints are illustrated in Table 2.2, which lists all of the quantities used to describe the machine behavior, with the fundamental dimensional quantities given in the right columns. These quantities can be expressed in any system of units. For example, one could choose the fundamental quantities as mass [M], length [L], time [T], and charge [Q] and express all quantities in terms of these [M-L-T-Q] units. If this is done, the dimensions of the common generator quantities are those given in column four of Table 2.2. Since most of the generator quantities are electrical, it is simpler to express them in terms of the fundamental electrical quantities of voltage [V], current [I], and time [T], as shown in column five of Table 2.2. Hereafter, we shall use the [VIT] system of units.

Table 2.2 Electrical Quantities, Units, and Dimensions

Quantity	Symbol	Units	Dimensions	
			M - L - T - Q	V - I - T
Voltage	v	volts	$ML^2T^{-2}Q^{-1}$	V
Current	i	amperes	$T^{-1}Q$	I
Voltamperes	S	voltamperes	ML^2T^{-3}	VI
Power	P	watts	ML^2T^{-3}	VI
Torque	T	newton-meters	ML^2T^{-2}	VIT
Flux Linkage	ψ	weber	$ML^2T^{-1}Q^{-1}$	VT
Resistance	r	ohm	$ML^2T^{-1}Q^{-2}$	VI^{-1}
Inductance	L or M	henry	ML^2Q^{-1}	$VI^{-1}T$
Time	t	second	T	T
Angular Velocity	ω	radian/second	T^{-1}	T^{-1}
Angle	θ	radian	---	---

One can show that it is essential to choose the same time base in all parts of the system (see [4, 9, and 10] for a discussion of this subject). Choosing a common time base forces the voltampere base to be equal in all parts of the system (e.g., the rotor and stator circuits) and forces the base mutual inductance to be the geometric mean of the base self inductances, if one is to obtain equal per unit mutuals. This is highly desirable. If we have equal per unit mutual inductances, then all off-diagonal mutual inductances in (2.64) are equal and, incidentally, the circuits are physically realizable.

Thus, in per unit, we may write

$$\begin{aligned} L_{AD} &= kM_F = kM_D = M_X \qquad \text{per unit} \\ L_{AQ} &= kM_G = kM_Q = M_Y \qquad \text{per unit.} \end{aligned} \tag{2.85}$$

We shall henceforth refer only to these per unit (pu) values. We also find it convenient to separate the machine self inductances into leakage and mutual terms. Thus, for the direct axis circuits we write

$$\begin{aligned} L_d &= \ell_a + L_{AD} \\ L_F &= \ell_F + L_{AD} \quad \text{pu} \\ L_D &= \ell_D + L_{AD} \end{aligned} \tag{2.86}$$

and for the quadrature axis

$$\begin{aligned} L_q &= \ell_a + L_{AQ} \\ L_G &= \ell_G + L_{AQ} \quad \text{pu} \\ L_Q &= \ell_Q + L_{AQ} \end{aligned} \tag{2.87}$$

where we define $\ell_a = \ell_d = \ell_q$. Then the normalized flux linkage equation may be written as

$$\begin{bmatrix} \psi_0 \\ \hline \psi_d \\ \psi_F \\ \psi_D \\ \hline \psi_q \\ \psi_G \\ \psi_Q \end{bmatrix} = \left[\begin{array}{c|ccc|ccc} L_0 & & & & & & \\ \hline & L_d & L_{AD} & L_{AD} & & & \\ & L_{AD} & L_F & L_{AD} & & & \\ & L_{AD} & L_{AD} & L_D & & & \\ \hline & & & & L_q & L_{AQ} & L_{AQ} \\ & & & & L_{AQ} & L_G & L_{AQ} \\ & & & & L_{AQ} & L_{AQ} & L_Q \end{array}\right] \begin{bmatrix} i_0 \\ \hline i_d \\ i_F \\ i_D \\ \hline i_q \\ i_G \\ i_Q \end{bmatrix} \tag{2.88}$$

where it is noted that the off diagonal terms of both 3 × 3 partitions are equal in both axes.

Synchronous machine operation under balanced three-phase conditions is of particular interest for SSR analysis. For this special case, we write (2.64) in normalized form as

$$\begin{bmatrix} v_d \\ -v_F \\ -v_D \\ \hline v_q \\ -v_G \\ -v_Q \end{bmatrix} = -\left[\begin{array}{ccc|ccc} r_a & & & & & \\ & r_F & & & & \\ & & r_D & & & \\ \hline & & & r_a & & \\ & & & & r_G & \\ & & & & & r_Q \end{array}\right] \begin{bmatrix} i_d \\ i_F \\ i_D \\ \hline i_q \\ i_G \\ i_Q \end{bmatrix}$$

$$-\frac{1}{\omega_B}\left[\begin{array}{ccc|ccc} L_d & L_{AD} & L_{AD} & & & \\ L_{AD} & L_F & L_{AD} & & & \\ L_{AD} & L_{AD} & L_D & & & \\ \hline & & & L_q & L_{AQ} & L_{AQ} \\ & & & L_{AQ} & L_G & L_{AQ} \\ & & & L_{AQ} & L_{AQ} & L_Q \end{array}\right] \begin{bmatrix} pi_d \\ pi_F \\ pi_D \\ \hline pi_q \\ pi_G \\ pi_Q \end{bmatrix} + \begin{bmatrix} -\omega\psi_q \\ 0 \\ 0 \\ \hline +\omega\psi_d \\ 0 \\ 0 \end{bmatrix} \text{pu} \tag{2.89}$$

where all quantities are in per unit except time, which is in seconds. Since time is in seconds the inverse base radian frequency multiplier to the derivative term is necessary. This equation can be more compactly written as

$$\begin{bmatrix} v_d \\ -v_F \\ -v_D \\ \hline v_q \\ -v_G \\ -v_Q \end{bmatrix} = -\left[\begin{array}{ccc|ccc} r_a & & & & & \\ & r_F & & & & \\ & & r_D & & & \\ \hline & & & r_a & & \\ & & & & r_G & \\ & & & & & r_Q \end{array}\right] \begin{bmatrix} i_d \\ i_F \\ i_D \\ \hline i_q \\ i_G \\ i_Q \end{bmatrix} - \frac{1}{\omega_B}\begin{bmatrix} p\psi_d \\ p\psi_F \\ p\psi_D \\ \hline p\psi_q \\ p\psi_G \\ p\psi_Q \end{bmatrix} + \begin{bmatrix} -\omega\psi_q \\ 0 \\ 0 \\ \hline +\omega\psi_d \\ 0 \\ 0 \end{bmatrix} \text{pu} \tag{2.90}$$

where the time variable is in seconds and the base radian frequency is in radians per second, but all other quantities are given in per unit.

The notation used here is common, and is the notation introduced by Kimbark [5], but it is arbitrary. There is no "standard" notation, and indeed standardization is not necessary. For example, one sometimes sees the mutual inductance defined as

$$L_{AD} = L_{md} \text{ or } L_{MD}. \tag{2.91}$$

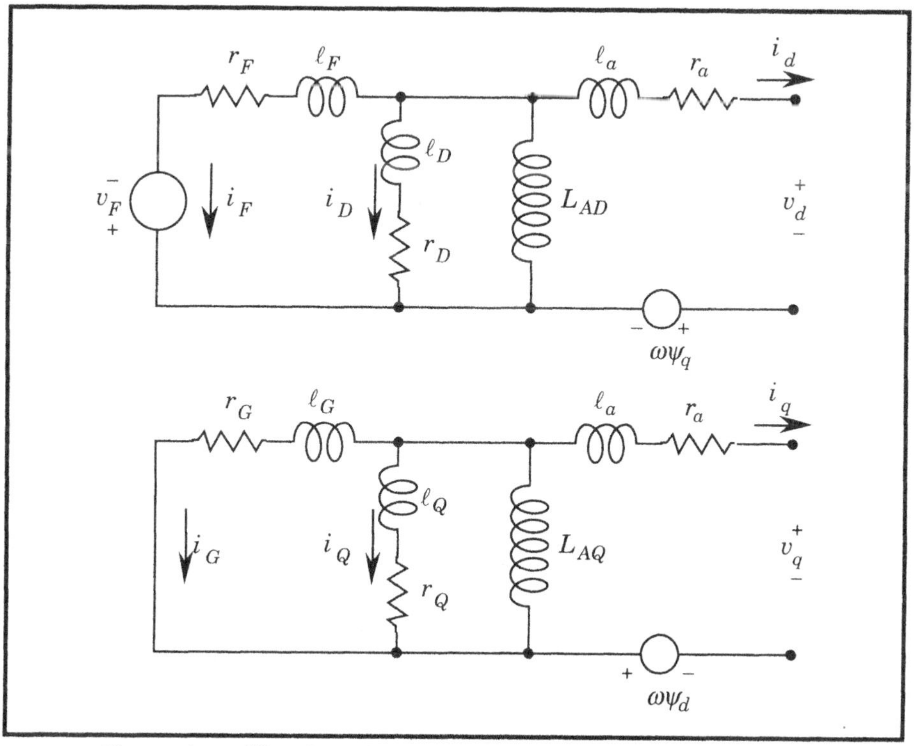

Figure 2.10 The d- and q-axis Per Unit Equivalent Circuits

The engineer who uses these equations will readily know the meaning of the terms, irrespective of the notation used. Care must be exercised in reading the many references, however, as there are differences that may be confusing.

We now write the per unit torque equation. From (2.81) and neglecting transient energy storage in the coupling fields, we write

$$T_{eg} = (\psi_d i_q - \psi_q i_d) \tag{2.92}$$

where we add the subscript "g" to emphasize that the torque is on the generator base. This will be discussed further below.

The per unit flux linkages are given by (2.88). We note that the flux linkage and voltage equations are all linear except for the speed voltage terms. The torque equation is nonlinear.

The machine circuit equations given by (2.89) and (2.90) are usually expressed schematically by the d and q equivalent circuits shown in Figure 2.10. These circuits are a simplified version of the circuits given in Figure 2.7. The simplification is possible since the mutual inductance is equal between all pairs of windings. This allows the construction of the tee equivalent to the mutually coupled circuits.

2.7 ANALYSIS OF THE DIRECT AXIS EQUATIONS

The generator equations given by (2.90) represent tightly coupled linear circuits in both the d- and q-axes, but with nonlinear controlled sources in each. Note, however, that the rotor equations in both axes, that is, those equations designated F-D and G-Q, are linear. These linear equations are described by rotor resistances and inductances that are not usually known in a physical machine. The parameters that *are* known are the transient and subtransient reactances and time constants of these circuits. We now develop the relationship between these known parameters and the equations given by (2.90). We do this by performing an analytical reduction of the equations. In performing this reduction, we recognize the linearity of the equations and make use of the Laplace Transform. We begin the analysis with the voltage equation for the F and D circuits, which we write in the Laplace domain as follows. First we write the F circuit voltage equation

$$v_F = r_F i_F + \frac{s\psi_F}{\omega_B} \tag{2.93}$$

and using the per unit flux linkage equation (2.88) for the field flux linkage term, we expand (2.93) to write

$$v_F = \left(r_F + \frac{sL_F}{\omega_B} \right) i_F + \left(\frac{sL_{AD}}{\omega_B} \right) i_d + \left(\frac{sL_{AD}}{\omega_B} \right) i_D . \tag{2.94}$$

In a similar way we compute the D circuit equation as

$$\begin{aligned} v_D = 0 &= r_D i_D + \frac{s\psi_D}{\omega_B} \\ &= \left(r_D + \frac{sL_D}{\omega_B} \right) i_D + \left(\frac{sL_{AD}}{\omega_B} \right) i_d + \left(\frac{sL_{AD}}{\omega_B} \right) i_F . \end{aligned} \tag{2.95}$$

We rewrite these equations in matrix form for clarity.

$$\begin{bmatrix} \left(r_F + \dfrac{sL_F}{\omega_B}\right) & \dfrac{sL_{AD}}{\omega_B} \\ \dfrac{sL_{AD}}{\omega_B} & \left(r_D + \dfrac{sL_D}{\omega_B}\right) \end{bmatrix} \begin{bmatrix} i_F \\ i_D \end{bmatrix} = \begin{bmatrix} v_F - \dfrac{sL_{AD}}{\omega_B} i_d \\ -\dfrac{sL_{AD}}{\omega_B} i_D \end{bmatrix}. \tag{2.96}$$

Solving (2.96) for the F and D currents, we have

$$\begin{bmatrix} i_F \\ i_D \end{bmatrix} = \frac{1}{\Delta} \begin{bmatrix} \left(r_D + \dfrac{sL_D}{\omega_B}\right) & -\dfrac{sL_{AD}}{\omega_B} \\ -\dfrac{sL_{AD}}{\omega_B} & \left(r_F + \dfrac{sL_F}{\omega_B}\right) \end{bmatrix} \begin{bmatrix} v_F - \dfrac{sL_{AD}}{\omega_B} i_d \\ -\dfrac{sL_{AD}}{\omega_B} i_d \end{bmatrix} \tag{2.97}$$

where Δ is the determinant of the coefficient matrix, or

$$\begin{aligned} \Delta &= \left(r_F + \frac{sL_F}{\omega_B}\right)\left(r_D + \frac{sL_D}{\omega_B}\right) - \frac{s^2 L_{AD}^2}{\omega_B^2} \\ &= \frac{L_F L_D - L_{AD}^2}{\omega_B^2} s^2 + \frac{r_F L_D + r_D L_F}{\omega_B} s + r_F r_D. \end{aligned} \tag{2.98}$$

Thus, we may write both the F and D currents in terms of the field voltage and the stator d-axis current. Now, from the flux linkage equation (2.88) we write

$$\psi_d = L_d\, i_d + L_{AD}\, i_F + L_{AD}\, i_D \tag{2.99}$$

and we recognize that currents of the second two terms may be substituted from (2.97). Substituting (2.97) into (2.99) we compute

$$\begin{aligned} \psi_d = L_d i_d + L_{AD}\left(r_D + \frac{sL_{AD}}{\omega_B}\right) v_F - \frac{sL_{AD}^2\left(r_D + \dfrac{sL_{AD}}{\omega_B}\right) i_d}{\omega_B \Delta} \\ + \frac{s^2 L_{AD}^3 i_d}{\omega_B^2 \Delta} - \frac{sL_{AD}^2 v_F}{\omega_B \Delta} + \frac{s^2 L_{AD}^3 i_d}{\omega_B^2 \Delta} - \frac{sL_{AD}^2\left(r_F + \dfrac{sL_F}{\omega_B}\right) i_d}{\omega_B \Delta} \end{aligned} \tag{2.100}$$

which can be simplified to

$$
\begin{aligned}
\psi_d &= \left[L_d - \frac{sL_{AD}^2\left(r_D + \dfrac{sL_D}{\omega_B}\right)}{\omega_B \Delta} + \frac{2s^2 L_{AD}^3}{\omega_B^2 \Delta} - \frac{sL_{AD}^2\left(r_F + \dfrac{sL_F}{\omega_B}\right)}{\omega_B \Delta} \right] i_d \\
&\quad + \left[L_{AD}\left(r_D + \frac{sL_D}{\omega_B}\right) - \frac{sL_{AD}^2}{\omega_B \Delta} \right] v_F \\
&= \left[\frac{b_2 s^2 + b_1 s + b_0}{a_2 s^2 + a_1 s + a_0} \right] i_d + \left[\frac{g_1 s + g_0}{a_2 s^2 + a_1 s + a_0} \right] v_F
\end{aligned}
\tag{2.101}
$$

where we define the coefficients

$$
\begin{aligned}
b_2 &= \frac{L_d L_F L_D - (L_d + L_F + L_D) L_{AD}^2 + 2L_{AD}^3}{\omega_B^2} \\
b_1 &= r_F\left(L_d L_D - L_{AD}^2\right) + r_F\left(L_d L_F - L_{AD}^2\right) \\
b_0 &= r_F r_D L_d \\
g_1 &= L_F L_{AD} - L_{AD}^2 \\
g_0 &= \omega_B r_F L_{AD} \\
a_2 &= \frac{L_F L_D - L_{AD}^2}{\omega_B^2} \\
a_1 &= \frac{r_F L_D + r_D L_F}{\omega_B} \\
a_0 &= r_F r_D .
\end{aligned}
$$

For convenience, we write (2.100) in the following form.

$$
\psi_d = L_d(s) i_d + G_d(s) v_F \tag{2.102}
$$

where the transfer functions L_d and G_d are defined by (2.101). We now seek an expanded definition of these transfer functions. The polynomial

Table 2.3 Derived d-axis Inductances and Time Constants (All inductances in per unit and all time constants in s)		
Name	Symbol	Definition
Subtransient Inductance	L_d''	$L_d - \dfrac{L_D L_{AD}^2 + L_F L_{AD}^2 - 2L_{AD}^3}{L_F L_D - L_{AD}^2}$
Transient Inductance	L_d'	$L_d - \dfrac{L_{AD}^2}{L_F}$
Subtransient Open Circuit Time Constant *	τ_{do}''	$\dfrac{L_F L_D - L_{AD}^2}{\omega_B r_D L_F}$
Transient Open Circuit Time Constant *	τ_{do}'	$\dfrac{L_F}{\omega_B r_F}$
Subtransient Short Circuit Time Constant *	τ_d''	$\dfrac{L_d''}{L_d'} \tau_{do}''$
Transient Short Circuit Time Constant *	τ_d'	$\dfrac{L_d'}{L_d} \tau_{do}'$
* ω_B is the system base radian frequency		

coefficients of (2.101) have patterns that appear in the definitions of transient and subtransient inductances and time constants. These important quantities are defined in Table 2.3.

One can show that the transfer function denominator may be written as

$$\Delta = r_F r_D \tau_{do}'' \tau_{do}' s^2 + \frac{r_F L_D + r_D L_F}{\omega_B} s + r_F r_D . \tag{2.103}$$

This expression could be factored if the s term were slightly different. Under closer examination, we may compute

$$r_F L_D + r_D L_F = \omega_B r_F r_D \left(\tau''_{do} + \tau'_{do} + \frac{L_{AD}^2}{r_D L_F \omega_B} \right). \tag{2.104}$$

Now, if we can show that

$$\tau'_{do} > \tau''_{do} >> \frac{L_{AD}^2}{r_D L_F \omega_B} \tag{2.105}$$

then we can write, approximately

$$r_F L_D + r_D L_F \approx r_F r_D \omega_B \left(\tau''_{do} + \tau'_{do} \right)$$

With this simplification, which may be a good approximation for some machines, we write (2.103) as

$$\begin{aligned} \Delta &= r_F r_D \left[\tau''_{do} \tau'_{do} s^2 + \left(\tau''_{do} + \tau'_{do} \right) s + 1 \right] \\ &= r_F r_D \left(1 + \tau'_{do} s \right) \left(1 + \tau''_{do} s \right) \end{aligned} \tag{2.106}$$

where the resistances are in per unit and the time constants are in seconds. Where the inequality (2.105) is not satisfied, the denominator must be left as a quadratic in s or factored into its two real roots.

The numerator parameters can be evaluated as follows. A straightforward expansion of the b factors will show that

$$\begin{aligned} b_2 &= \frac{\left(L_F L_D - L_{AD}^2 \right) L''_d}{\omega_B} \\ &= \omega_B^2 r_F r_D \tau''_{do} \tau'_{do} L''_d \end{aligned} \tag{2.107}$$

Also

$$\begin{aligned} b_1 &= \omega_B r_F r_D \left[\frac{L_D}{\omega_B r_D} \left(L_d - \frac{L_{AD}^2}{L_D} \right) + \frac{L_F}{\omega_B r_F} \left(L_d - \frac{L_{AD}^2}{L_F} \right) \right] \\ &= \omega_B r_F r_D \left(\frac{\tau''_{do} L_d L''_d}{L'_d} + \tau'_{do} L'_d \right) \\ &= \omega_B r_F r_D \left(\tau''_d + \tau'_d \right) L_d \end{aligned} \tag{2.108}$$

where we use the definition of short circuit time constants from Table 2.3.

Finally, factoring the numerator polynomial, we compute

$$b_2s^2 + b_1s + b_0 = \omega_B r_F r_D (1+\tau''_d s)(1+\tau'_d s)L_d \tag{2.109}$$

and we write the first transfer function as

$$L_d(s) = \frac{L_d(1+\tau''_d s)(1+\tau'_d s)}{(1+\tau''_{do}s)(1+\tau'_{do}s)}. \tag{2.110}$$

In a similar manner, we compute

$$G_d(s) = \frac{K_d(1+\tau'_{Fo}s)}{(1+\tau''_{do}s)(1+\tau'_{do}s)} \tag{2.111}$$

where we have defined the following gain and time constant

$$K_d = \frac{\omega_B \tau''_{do}(L_d - \ell_a)(L'_d - L''_d)}{(L'_d - \ell_a)^2} \tag{2.112}$$

and

$$\tau'_{Fo} = \tau'_{do}\frac{(L'_d - \ell_a)}{(L_d - \ell_a)} \tag{2.113}$$

Finally, then we may write

$$\begin{aligned}\psi_d(s) &= L_d(s)i_d + G_d(s)v_F \\ &= \frac{L_d(1+\tau''_d s)(1+\tau'_d s)}{(1+\tau''_{do}s)(1+\tau'_{do}s)}i_d(s) + \frac{K_d(1+\tau'_{Fo}s)}{(1+\tau''_{do}s)(1+\tau'_{do}s)}v_F(s).\end{aligned} \tag{2.114}$$

The importance of these transfer functions is that the data are generally in the form supplied by the manufacturer. This is certainly true of (2.110), which we will find is an important function in SSR analysis.

2.8 ANALYSIS OF THE QUADRATURE AXIS EQUATIONS

The quadrature axis equations are analyzed in exactly the same way as the direct axis equations. This is readily accomplished by replacing all d-axis subscripts by the corresponding q-axis subscripts, since the two networks are identical in form. The resulting equations are as follows:

$$\psi_q = L_q(s)\, i_q + G_q(s)\, v_G \tag{2.115}$$

where

$$L_q(s) = \frac{L_q\left(1+\tau_q''s\right)\left(1+\tau_q's\right)}{\left(1+\tau_{qo}''s\right)\left(1+\tau_{qo}'s\right)} \tag{2.116}$$

$$G_q(s) = \frac{K_q\left(1+\tau_{Go}'s\right)}{\left(1+\tau_{qo}''s\right)\left(1+\tau_{qo}'s\right)}. \tag{2.117}$$

Most machines have only d-axis excitation, in which case we can set the G excitation voltage to zero.

Then, for $v_G = 0$

$$\psi_q(s) = \frac{L_q\left(1+\tau_q''s\right)\left(1+\tau_q's\right)}{\left(1+\tau_{qo}''s\right)\left(1+\tau_{qo}'s\right)} i_q(s). \tag{2.118}$$

2.9 SUMMARY OF MACHINE EQUATIONS

We may summarize the synchronous machine equations as follows:

(1) The circuit differential equations, from (2.90):

$$v_d = -r_a i_d - \frac{1}{\omega_B}\frac{d\psi_d}{dt} - \omega\psi_q$$

$$v_q = -r_a i_q - \frac{1}{\omega_B}\frac{d\psi_q}{dt} + \omega\psi_d. \tag{2.119}$$

(2) The rotor transfer function equations from (2.102) and (2.115):

$$\begin{aligned}\psi_d(s) &= L_d(s)i_d + G_d(s)v_F\\ \psi_q(s) &= L_q(s)i_q + G_q(s)v_G.\end{aligned} \tag{2.120}$$

(3) The speed and torque equations from Newton's Law and (2.92), expressed on a system base rather than the machine base:

$$\begin{aligned}\omega(s) &= \frac{1}{2Hs}\left(T_m - T_e - D\Delta\omega\right)\\ T_e &= \frac{1}{3}\left(\psi_d i_q - \psi_q i_d\right).\end{aligned} \tag{2.121}$$

The parameter H in (2.128), called the inertia constant, is discussed in Chapter 4 and is defined by (4.27). The one-third constant multiplying the electromagnetic torque term is for a change of base, and is explained in the next section.

These equations are a mixture of time domain equations, for the nonlinear relationships, and Laplace domain equations for the linear relationships.

It is sometimes convenient to write the swing equation in terms of power rather than torque, since the mechanical output of the turbine is often given in terms of power. To convert torque to power, we write for any electrical power

$$P_e = T_e\,\omega = P_{eo} + T_{eo}\,\Delta\omega + \omega_o\Delta T_e \tag{2.122}$$

We also usually modify the swing equation to compute only the *change* in speed $\Delta\omega$. Thus (2.121) is written as

$$\Delta\omega(s) = \frac{1}{2Hs}\left(\Delta T_m - \Delta T_e - D\Delta\omega\right) \quad \text{pu} \tag{2.123}$$

where all quantities are in per unit on the system base.

We now address the problem of converting the generator quantities from the generator base to the system base, and vice versa.

2.10 MACHINE - NETWORK INTERFACE EQUATIONS

Having developed the normalized equations of the machine, it is necessary to interface these equations with the electric network, the excitation system, and the swing equation. Figure 1.3 provides an overview of the system and the generator's place in the system. The generator block of Figure 1.3 is repeated here as Figure 2.11.

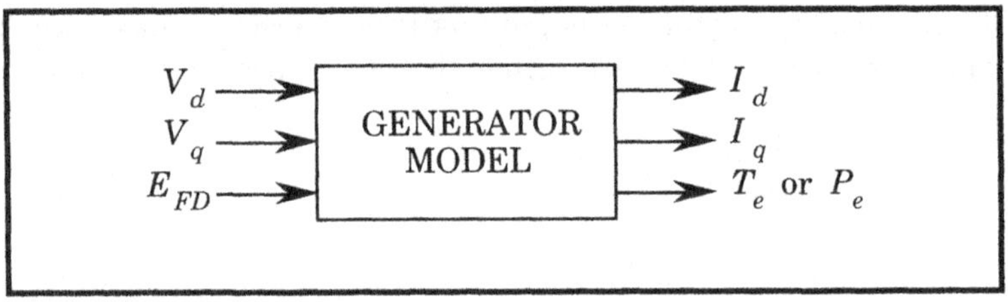

Figure 2.11 Block Diagram of the Generator Model

Clearly, the generator model has three inputs – the two voltage components, which can be derived from the phasor terminal voltage in the network model, and the field voltage from the excitation system. In SSR studies, the excitation system is not always represented, in which case the field voltage is taken to be a constant and this part of the model is ignored in the SSR calculations.

Although the representation shown in Figure 2.11 is entirely proper, there remains one last step in interfacing the machine model with the input and output quantities shown. This last step has to do with the base quantities used in the normalization. In the network, the normalization is performed using different base quantities than those chosen for the generator normalization. Actually, the choice of base quantities is arbitrary and several different schemes are common. Table 2.4 provides a comparison of network base quantities and the machine base quantities used in this book.

Clearly, a change of base is required in moving from the network equations (in per unit) to the generator equations, and vice versa.

In practice, this is a simple change of base that is easily accomplished. Graphically, the block diagram of this change is shown in Figure 2.12.

Table 2.4 Comparison of Network and Machine Base Quantities		
Quantity	Network Base	Generator Base
Voltampere	Arbitrary	Machine Rated VA per Phase
Voltage	Rated Line-to-Line	Machine Rated Line-to-Neutral

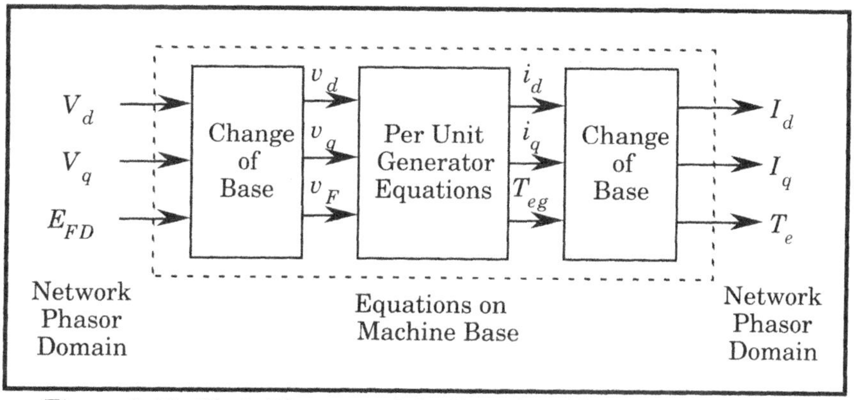

Figure 2.12 Block Diagram of the Interface Showing the Change of Base Required for Correct Results

Actually, it would be possible to change the generator equations to the same base as that of the network. However, this destroys one of the major advantages of the per unit system, namely, the familiarity of the per unit quantities. In many respects, it is preferable to leave the generator equations on their own base, using data exactly as provided by the manufacturer. This is the approach that will be used here, since it allows generator parameters that are recognizable as to their magnitude, and gives the user the benefit of a quick visual check of their accuracy.

The base change that is needed is shown in Table 2.5. There are two concepts at work in the base conversion. The network is always normalized on the basis of an arbitrary three-phase voltampere base and the nominal rated line-to-line voltages of the network branches as the base voltages.

Table 2.5 Change of Base Between System and Generator Quantities, where All Quantities Shown are in Per Unit

Transformation	Quantity	Base Change Multiplier	Output Quantity
System to Generator	V_d	$\sqrt{3}$	v_d
	V_d	$\sqrt{3}$	v_q
	E_{FD}	$\dfrac{\sqrt{3}\,r_F}{\omega_B L_{AD}}$	v_F
Generator to System	i_d	$\dfrac{1}{\sqrt{3}}\dfrac{S_{GR}}{S_B}\dfrac{V_{B(LL)}}{V_{R(LL)}}$	I_d
	i_q	$\dfrac{1}{\sqrt{3}}\dfrac{S_{GR}}{S_B}\dfrac{V_{B(LL)}}{V_{R(LL)}}$	I_q
	T_{eg}	$\dfrac{1}{3}\dfrac{S_B}{S_{GR}}$	T_e

Legend of Multiplier Quantities

S_{GR} = Rated Generator 3 Phase MVA

S_B = Arbitrary Network 3 Phase MVA Base

$V_{B(LL)}$ = Network Base Line- to - Line Voltage

$V_{R(LL)}$ = Machine Rated Line- to - Line Voltage

These are not the base voltampere and base voltage used in the machine equations. The machine variables must be converted to quantities based on the machine rated line-to-line voltage and three-phase rated MVA. However, these rated generator quantities will seldom agree with the system base MVA and base voltage, so a second conversion is necessary for this change of base, thereby expressing all generator outputs on the system base.

The generator input conversion is relatively easy since it involves only the per unit voltage, which is a per unit voltage on any base. The d- and q-axis network voltages are stator equivalent per unit rms quantities that are taken from the network phasor voltage representation. These need to be scaled to the machine base. The exciter voltage is already normalized to the machine and needs to be scaled as shown in Table 2.5 to agree with the machine normalization scheme.

2.11 LINEAR STATE-SPACE MACHINE EQUATIONS

Some of the machine equations are nonlinear and must be linearized for eigenvalue computation. We write all the generator equations in incremental form, using the "*o*" subscript to indicate the initial condition. The nonlinear equations can thus be written as follows.

$$\begin{aligned}
\Delta v_d &= -r_a\,\Delta i_d - \frac{1}{\omega_B}\frac{d\Delta\psi_d}{dt} - \omega_o\,\Delta\psi_q - \psi_{qo}\,\Delta\omega \\
\Delta v_q &= -r_a\,\Delta i_q - \frac{1}{\omega_B}\frac{d\Delta\psi_q}{dt} + \omega_o\,\Delta\psi_d + \psi_{do}\,\Delta\omega \\
\Delta T_{eg} &= \psi_{do}\,\Delta i_q + i_{qo}\,\Delta\psi_d - \psi_{qo}\,\Delta i_d - i_{do}\,\Delta\psi_q .
\end{aligned} \tag{2.124}$$

The remaining linear equations are written by simply replacing all variables by the incremental (Δ) variables.

Now, since *all* variables are incremental, we drop the Δ notation for simplicity and write all equations in the s domain. First, we have the differential equations

$$\begin{aligned}
\frac{s\psi_d}{\omega_B} &= -v_d - r_a i_d - \omega_o\psi_q - \psi_{qo}\omega \\
\frac{s\psi_q}{\omega_B} &= -v_q - r_a i_q + \omega_o\psi_d + \psi_{do}\omega \\
2Hs\omega &= T_m - T_e - D\omega
\end{aligned} \tag{2.125}$$

where all quantities are in per unit on the system base except s, which has the dimensions of 1/seconds (s^{-1}).

These equations are supported by the auxiliary equations, assuming only a d-axis field winding,

$$\begin{aligned}
\psi_d(s) &= L_d(s)i_d(s) + G_d(s)v_F(s) \\
\psi_q(s) &= L_q(s)i_q(s) + G_q(s)v_G(s)
\end{aligned}$$

$$T_e = \frac{T_{eg}}{3} = \frac{\left(\psi_{do} i_q + i_{qo}\psi_d - \psi_{qo} i_d - i_{do}\psi_q\right)}{3}. \tag{2.126}$$

These equations are a mixture of Laplace and time domain equations. They represent a solution of the machine equations that may be represented in block diagram form, as shown in Figure 2.13. This is a convenient form for solutions using certain types of computer programs that can handle the mixture of Laplace and time domain expressions.

To write all the equations as state-space equations, they must all be in the time domain and must all be linear. The equations (2.125) and (2.126) are linear, but are not in a very convenient form.

We now arrange these equations in the standard linear form

$$\dot{\mathbf{Y}}_1 = \mathbf{A}\mathbf{Y}_1 + \mathbf{B}\mathbf{F} \tag{2.127}$$

where

$\mathbf{Y}_1$ = the n vector of state variables
$\mathbf{F}$ = the m vector of input variables
$\mathbf{A}$ = an n x n matrix of constants
$\mathbf{B}$ = an n x m matrix of constants

We can do this by returning to the flux linkage equation (2.88), which we write in matrix form as

$$\boldsymbol{\psi} = \mathbf{L}\,\mathbf{i}. \tag{2.128}$$

Since the $\mathbf{L}$ matrix is nonsingular, we solve for the current $\mathbf{i}$ to write

$$\mathbf{i} = \mathbf{L}^{-1}\boldsymbol{\psi} = \boldsymbol{\Gamma}\boldsymbol{\psi} \tag{2.129}$$

where we compute the matrix $\boldsymbol{\Gamma}$ as

$$\boldsymbol{\Gamma} = \left[\begin{array}{ccc|ccc} \Gamma_{dd} & \Gamma_{dF} & \Gamma_{dD} & & & \\ \Gamma_{Fd} & \Gamma_{FF} & \Gamma_{FD} & & & \\ \Gamma_{Dd} & \Gamma_{DF} & \Gamma_{DD} & & & \\ \hline & & & \Gamma_{qq} & \Gamma_{qG} & \Gamma_{qQ} \\ & & & \Gamma_{Gq} & \Gamma_{GG} & \Gamma_{GQ} \\ & & & \Gamma_{Qq} & \Gamma_{QG} & \Gamma_{QQ} \end{array}\right] \tag{2.130}$$

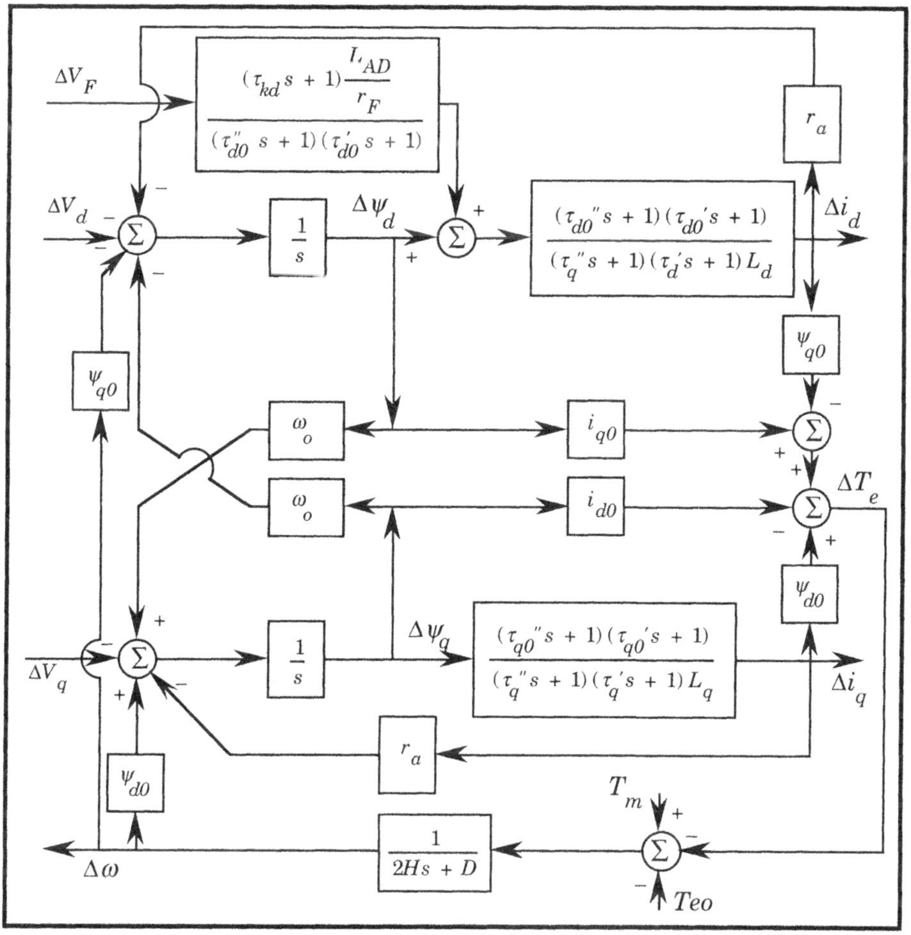

Figure 2.13 Block Diagram of the Linear Generator Equations

and it can be shown that this matrix is symmetric [4]. Returning to (2.90), we rearrange the terms of the equation to write

$$\begin{bmatrix} \dot\psi_d \\ \dot\psi_F \\ \dot\psi_D \\ \hline \dot\psi_q \\ \dot\psi_G \\ \dot\psi_Q \end{bmatrix} = \left[\begin{array}{ccc|ccc} r_a & & & & & \\ & r_F & & & & \\ & & r_D & & & \\ \hline & & & r_a & & \\ & & & & r_G & \\ & & & & & r_Q \end{array}\right] \begin{bmatrix} i_d \\ i_F \\ i_D \\ \hline i_q \\ i_G \\ i_Q \end{bmatrix} + \begin{bmatrix} -\omega_B \omega \psi_q \\ 0 \\ 0 \\ \hline +\omega_B \omega \psi_d \\ 0 \\ 0 \end{bmatrix} + \begin{bmatrix} -\omega_B v_d \\ \omega_B v_F \\ 0 \\ \hline -\omega_B v_q \\ 0 \\ 0 \end{bmatrix}. \tag{2.131}$$

If we substitute (2.129) for the current vector, then (2.131) will be written in terms of only flux linkages as state variables. The result may be written as

$$\begin{bmatrix} \dot{\psi}_d \\ \dot{\psi}_F \\ \dot{\psi}_D \\ \hline \dot{\psi}_q \\ \dot{\psi}_G \\ \dot{\psi}_Q \end{bmatrix} = \left[\begin{array}{ccc|ccc} h_{dd} & h_{dF} & h_{dD} & & & \\ h_{Fd} & h_{FF} & h_{FD} & & & \\ h_{Dd} & h_{DF} & h_{DD} & & & \\ \hline & & & h_{qq} & h_{qG} & h_{qQ} \\ & & & h_{Gq} & h_{GG} & h_{GQ} \\ & & & h_{Qq} & h_{QG} & h_{QQ} \end{array}\right] \begin{bmatrix} \psi_d \\ \psi_F \\ \psi_D \\ \psi_q \\ \psi_G \\ \psi_Q \end{bmatrix} + \begin{bmatrix} -\omega_B \omega \psi_q \\ 0 \\ 0 \\ \hline +\omega_B \omega \psi_d \\ 0 \\ 0 \end{bmatrix} + \begin{bmatrix} -\omega_B v_d \\ \omega_B v_F \\ 0 \\ \hline -\omega_B v_q \\ 0 \\ 0 \end{bmatrix}. \tag{2.132}$$

where we have defined new constant coefficients of the form

$$h_{ik} = \omega_B r_i \Gamma_{ik} \tag{2.133}$$

where the index i refers to the element row and the index k refers to the column. These matrix elements are not symmetric, but they are all real. Note that the dimensions of h are derived to be s^{-1}.

Equation (2.132) includes the nonlinear speed voltage terms, which we linearize as follows. Symbolically, we let one of these terms be written as

$$v = \omega \psi_q$$

$$\Delta v = \omega_o \Delta \psi_q + \psi_{qo} \Delta \omega \tag{2.134}$$

and similarly for the q-axis term.

The other nonlinearity in the system equations is in the electromagnetic torque equation

$$T_{eg} = \psi_d i_q - \psi_q i_d. \tag{2.135}$$

For small deviations of the variables, we write

$$\Delta T_{eg} = \psi_{do} \Delta i_q + i_{qo} \Delta \psi_d - \psi_{qo} \Delta i_d - i_{do} \Delta \psi_q. \tag{2.136}$$

But the currents in (2.136) can be written in terms of the flux linkages using (2.129), with the result

$$
\begin{aligned}
i_d &= \Gamma_{dd}\psi_d + \Gamma_{dF}\psi_F + \Gamma_{dD}\psi_D \\
i_q &= \Gamma_{qq}\psi_q + \Gamma_{qG}\psi_G + \Gamma_{qQ}\psi_Q.
\end{aligned}
\tag{2.137}
$$

Substituting these currents into (2.136) and eliminating the Δ's for simplicity, we write

$$
\begin{aligned}
T_{eg} = {} & (\psi_{do}\Gamma_{qq} - i_{do})\psi_q + \psi_{do}\Gamma_{qG}\psi_G + \psi_{do}\Gamma_{qQ}\psi_Q \\
& - (\psi_{qo}\Gamma_{dd} - i_{qo})\psi_d - \psi_{qo}\Gamma_{dF}\psi_F - \psi_{qo}\Gamma_{dD}\psi_D
\end{aligned}
\tag{2.138}
$$

where all coefficients have the dimensions of per unit current.

The electromagnetic torque is required in the linearized swing equation given by (2.125), which requires the change of base given in Table 2.5 that multiplies (2.138) by 1/3. We simplify the notation of the result to write, in more compact form,

$$
T_e = I_{qq}\psi_q + I_{qG}\psi_G + I_{qQ}\psi_Q - I_{dd}\psi_d - I_{dF}\psi_F - I_{dD}\psi_D
\tag{2.139}
$$

where the change of base factor has been included in the new defined constants. This new equation may be substituted into the swing equation, which expresses this state equation in terms of the same state variables as used for the generator circuits. The result is a system of seven differential equations that may be written as follows.

$$
\left[\begin{array}{c}
\dot\psi_d \\ \dot\psi_F \\ \dot\psi_D \\ \hline \dot\psi_q \\ \dot\psi_G \\ \dot\psi_Q \\ \hline \dot\omega
\end{array}\right]
=
\left[\begin{array}{ccc|ccc|c}
h_{dd} & h_{dF} & h_{dD} & -\omega_B\omega_o & & & -\omega_B\psi_o \\
h_{Fd} & h_{FF} & h_{FD} & & & & \\
h_{Dd} & h_{DF} & h_{DD} & & & & \\ \hline
\omega_B\omega_o & & & h_{qq} & h_{qG} & h_{qQ} & \omega_B\psi_o \\
 & & & h_{Gq} & h_{GG} & h_{GQ} & \\
 & & & h_{Qq} & h_{QG} & h_{QQ} & \\ \hline
\dfrac{I_{dd}}{2H} & \dfrac{I_{dF}}{2H} & \dfrac{I_{dD}}{2H} & \dfrac{-I_{qq}}{2H} & \dfrac{-I_{qG}}{2H} & \dfrac{-I_{qQ}}{2H} & \dfrac{-D}{2H}
\end{array}\right]
\left[\begin{array}{c}
\psi_d \\ \psi_F \\ \psi_D \\ \hline \psi_q \\ \psi_G \\ \psi_Q \\ \hline \omega
\end{array}\right]
+
\left[\begin{array}{c}
-\omega_B v_d \\ \omega_B v_F \\ 0 \\ \hline -\omega_B v_q \\ 0 \\ 0 \\ \hline \dfrac{T_m}{2H}
\end{array}\right].
\tag{2.140}
$$

This is the desired state-space form for the generator equations. Other forms are possible and may be preferred in some cases [4]. The **A** matrix in (2.140) is clearly identified. The equation could be written in a somewhat different form to identify the **B** matrix, with the four variables on the right as control variables or inputs. This is left as an exercise for the interested reader, since the extension is quite straightforward.

2.12 EXCITATION SYSTEMS

In many cases, the analysis of SSR does not require the modeling of the excitation system, but in some cases the analyst may wish to study the effects of excitation. This section presents a simple extension of the previous work to show how the excitation system may be added to the machine equations.

Excitation systems are of many different designs, and there is no single mathematical model that is adequate for all types. Linear models for several types are given in [4]. We illustrate the way in which the excitation system modifies the machine equations by way of an example. Consider the excitation system shown in Figure 2.14, which represents a thyristor exciter with terminal voltage supply [4]. The linear equations for this excitation system may be written directly from the block diagram, assuming that the field voltage E_{FD} is operating between the limits. Thus we write

$$\begin{aligned}
\dot{V}_1 &= \frac{K_R}{\tau_R} V_t - \frac{1}{\tau_R} V_1 \\
\dot{V}_3 &= \frac{K_F}{\tau_F} \dot{E}_{FD} - \frac{1}{\tau_F} V_3 \\
\dot{E}_{FD} &= \frac{K_A}{\tau_A}\left(V_{REF} + V_S - V_1 - V_3\right) - \frac{1}{\tau_A} E_{FD}.
\end{aligned} \tag{2.141}$$

These equations are not in state space form due to the derivative term on the right side of the second equation. They are readily converted to the desired form by substituting the third equation into the second, with the result

$$\begin{bmatrix} \dot{V}_1 \\ \dot{V}_3 \\ \dot{E}_{FD} \end{bmatrix} = \begin{bmatrix} A_{88} & 0 & 0 \\ A_{98} & A_{99} & 0 \\ A_{10-8} & A_{10-9} & A_{10-10} \end{bmatrix} \begin{bmatrix} V_1 \\ V_3 \\ E_{FD} \end{bmatrix}$$

$$+ \begin{bmatrix} B_{81} & B_{82} & 0 & 0 \\ 0 & 0 & B_{94} & B_{95} \\ 0 & 0 & B_{10-4} & B_{10-5} \end{bmatrix} \begin{bmatrix} v_d \\ v_q \\ V_{REF} \\ V_S \end{bmatrix} \tag{2.142}$$

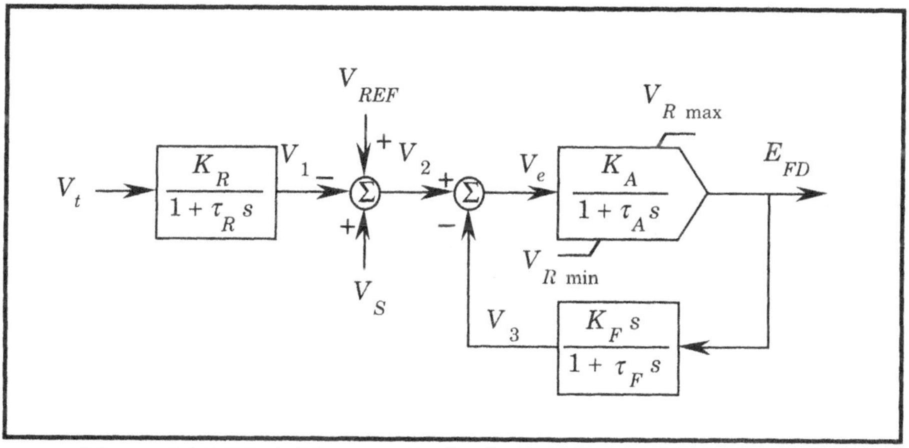

Figure 2.14 Block Diagram for Excitation System Type 1S [11].

where

$$
\begin{aligned}
A_{88} &= \frac{-1}{\tau_R} \\
A_{98} &= \frac{-K_F K_A}{\tau_F \tau_A} & B_{81} &= \frac{v_{do} K_R}{V_{to} \tau_R} \\
A_{99} &= \frac{-1}{\tau_F}\left(\frac{K_F K_A}{\tau_A} + 1\right) & B_{82} &= \frac{v_{qo} K_R}{V_{to} \tau_R}
\end{aligned} \qquad (2.143a)
$$

$$
\begin{aligned}
A_{9-10} &= \frac{-K_F}{\tau_F \tau_A} & B_{94} &= \frac{K_F K_A}{\tau_F \tau_A} \\
A_{10-8} &= \frac{-K_A}{\tau_A} & B_{95} &= B_{94} \\
A_{10-9} &= A_{10-8} & B_{10-4} &= \frac{K_A}{\tau_A} \\
A_{10-10} &= \frac{-1}{\tau_A} & B_{10-5} &= B_{10-4}.
\end{aligned} \qquad (2.143b)
$$

These equations are now combined with those for the generator, given by (2.140). This enlarges the system representation from 7th order to 10th order by adding the three new state variables given by (2.142). The resulting new state-space equation is given below. Note that the exciter voltage reference and the power system stabilizer output V_S now become input variables. The power system stabilizer could easily be added to the model. A speed input stabilizer would utilize the rotor speed as its input and its output would be injected into the voltage regulator summing junction, thereby closing the loop. This would add several more states to the model.

$$
\begin{bmatrix} \dot{\psi}_d \\ \dot{\psi}_F \\ \dot{\psi}_D \\ \hline \dot{\psi}_q \\ \dot{\psi}_G \\ \dot{\psi}_Q \\ \hline \dot{\omega} \\ \hline \dot{V}_1 \\ \dot{V}_3 \\ \dot{E}_{FD} \end{bmatrix}
=
\left[\begin{array}{ccc|ccc|c|ccc}
h_{dd} & h_{dF} & h_{dD} & -\omega_B\omega_o & 0 & 0 & -\omega_B\psi_{qo} & 0 & 0 & 0 \\
h_{Fd} & h_{FF} & h_{FD} & 0 & 0 & 0 & 0 & 0 & 0 & k_x \\
h_{Dd} & h_{DF} & h_{DD} & 0 & 0 & 0 & 0 & 0 & 0 & 0 \\
\hline
\omega_B\omega_o & 0 & 0 & h_{qq} & h_{qG} & h_{qQ} & +\omega_B\psi_{do} & 0 & 0 & 0 \\
0 & 0 & 0 & h_{Gq} & h_{GG} & h_{GQ} & 0 & 0 & 0 & 0 \\
0 & 0 & 0 & h_{Qq} & h_{QG} & h_{QQ} & 0 & 0 & 0 & 0 \\
\hline
\frac{I_{dd}}{2H} & \frac{I_{dF}}{2H} & \frac{I_{dD}}{2H} & \frac{-I_{qq}}{2H} & \frac{-I_{qG}}{2H} & \frac{-I_{qQ}}{2H} & \frac{-D}{2H} & 0 & 0 & 0 \\
\hline
0 & 0 & 0 & 0 & 0 & 0 & 0 & A_{88} & 0 & 0 \\
0 & 0 & 0 & 0 & 0 & 0 & 0 & A_{98} & A_{99} & 0 \\
0 & 0 & 0 & 0 & 0 & 0 & 0 & A_{10-8} & A_{10-9} & A_{10-10}
\end{array}\right]
\begin{bmatrix} \psi_d \\ \psi_F \\ \psi_D \\ \hline \psi_q \\ \psi_G \\ \psi_Q \\ \hline \omega \\ \hline V_1 \\ V_3 \\ E_{FD} \end{bmatrix}
$$

$$
+\left[\begin{array}{ccccc}
-\omega_B & 0 & 0 & 0 & 0 \\
0 & 0 & 0 & 0 & 0 \\
0 & 0 & 0 & 0 & 0 \\
\hline
0 & -\omega_B & 0 & 0 & 0 \\
0 & 0 & 0 & 0 & 0 \\
0 & 0 & 0 & 0 & 0 \\
\hline
0 & 0 & \frac{1}{2H} & 0 & 0 \\
\hline
B_{81} & B_{82} & 0 & 0 & 0 \\
0 & 0 & 0 & B_{94} & B_{95} \\
0 & 0 & 0 & B_{10-4} & B_{10-5}
\end{array}\right]
\begin{bmatrix} v_d \\ v_q \\ T_m \\ V_{REF} \\ V_S \end{bmatrix}
\tag{2.144}
$$

where

$$k_x = \frac{\sqrt{3}\, r_F}{L_{AD}}$$

takes care of the change of base between exciter and generator. With the equations in this form, one can easily identify the **A** and **B** matrices.

2.13 SYNCHRONOUS MACHINE SATURATION

Synchronous machine saturation is often defined in reference to the machine open-circuited saturation curve, shown in Figure 2.14.

When the open-circuited generator is running at synchronous speed and with balanced voltages in the three phases, we may write

$$\mathbf{v}_{abc} = \mathbf{P}^{-1}\mathbf{v}_{0dq}. \tag{2.145}$$

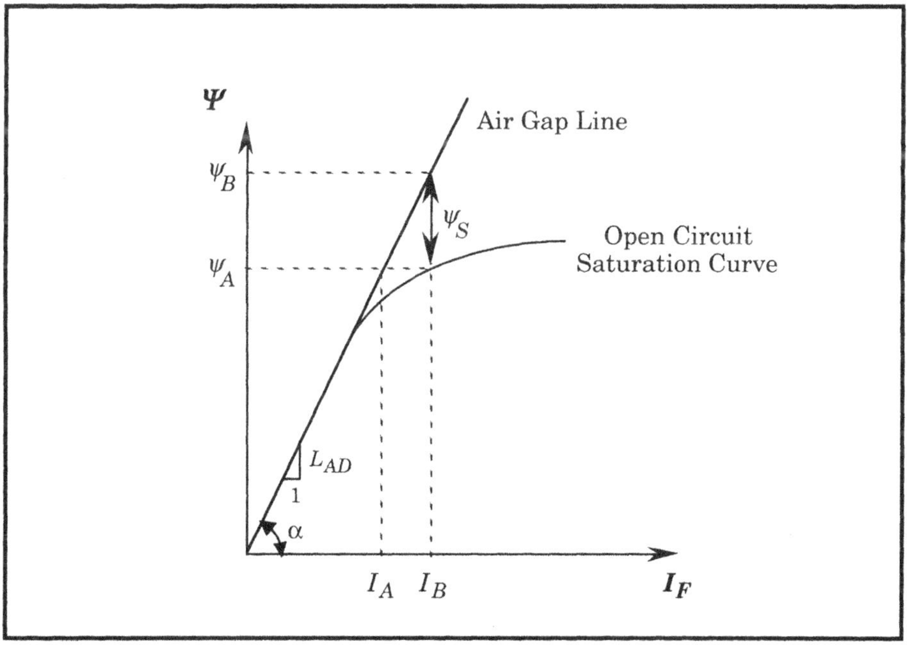

Figure 2.14 Synchronous Machine Open-Circuited Saturation Curve

Since the voltages are balanced, we know that the zero sequence voltage is

$$v_0 = 0 \tag{2.146}$$

and the phase "a" voltages may be written from (2.135) as

$$v_a = \sqrt{\frac{2}{3}}\left(v_d \cos\theta + v_q \sin\theta\right) \tag{2.147}$$

where

$$\theta = \omega_B t + \delta + \frac{\pi}{2} \tag{2.148}$$

Since the armature currents are zero and the speed is exactly one per unit, the d- and q-axis voltages are given from Figure 2.10 as

$$v_d = 0$$
$$v_q = \omega\psi_d = \omega L_{AD}\, i_F = L_{AD}\, i_F \tag{2.149}$$

where the final result is true since the speed is unity.

Then (2.137) becomes

$$\begin{aligned} v_a &= \sqrt{\frac{2}{3}}\, L_{AD}\, i_F \sin\theta \\ &= \sqrt{\frac{2}{3}}\, L_{AD}\, i_F \cos(\omega_B t + \delta) \\ &= \sqrt{2}\, E \cos(\omega_B t + \delta) \end{aligned} \tag{2.150}$$

where we define the rms open-circuit phase voltage to be

$$E = \frac{L_{AD}\, i_F}{\sqrt{3}} = L_{AD}\, I_F \tag{2.151}$$

where we carefully note the difference between i_F and I_F. We may think of the open-circuit voltage as being a function of the "stator equivalent" field current, defined as

$$I_F = \frac{i_F}{\sqrt{3}} \tag{2.152}$$

and from (2.151) we recognize the equation of a straight line through the origin with slope L_{AD}.

Since the product of inductance and current is a flux linkage, the voltage and flux linkage in per unit are exactly equal, or

$$V_a = E = \psi \quad \text{pu.} \tag{2.153}$$

We also observe the following:

1. If there is no saturation, the open-circuit voltage will vary from ψ_A to ψ_B in Figure 2.14 as the field current I_F changes from I_A to I_B.

2. The slope of the air gap line is exactly equal to the per unit mutual inductance L_{AD}.

3. As more current flows in the mutual inductance, saturation occurs such that at current I_B we have voltage ψ_A rather than ψ_B.

By definition, then, we write the d-axis saturation function as

$$S_{GD} = \frac{I_B - I_A}{I_A}. \tag{2.154}$$

For generators, the saturation function is usually provided at two values of ψ corresponding to 1.0 and 1.2 per unit open-circuit voltage. Then we have

$$S_{G1.0} = \frac{I_B - I_A}{I_A}$$

$$S_{G1.2} = \frac{I_C - 1.2I_A}{1.2I_A} \tag{2.155}$$

where

I_A = Current Corresponding to $\psi_A = 1.0$ on the Air Gap Line

I_B = Current Corresponding to $\psi_A = 1.0$ on the Saturation Curve

I_C = Current Corresponding to $\psi_C = 1.2$ on the Saturation Curve.

The saturation function is obviously a nonlinear function of voltage or field current that is zero for small values of the independent variable. Usually we use voltage as the independent variable and assume that saturation is negligible for voltages smaller than about 0.8 per unit, or mathematically

$$S_{GD} = 0 \quad \text{for} \quad \psi < 0.8 \text{ per unit.} \tag{2.156}$$

Many nonlinear functions can be used for this purpose, but two are in common usage. These are defined as follows:

1. <u>Exponential Saturation Fraction</u>

Define

$$S_{GD} = \begin{cases} A_G \exp\left[B_G(\psi - 0.8)\right], & \psi \geq 0.8 \\ 0, & \psi < 0.8. \end{cases} \tag{2.157}$$

2. Quadratic Saturation Function

Define

$$S_{GD} = \begin{cases} \dfrac{B_G(\psi - 0.8)^2}{\psi}, & \psi \geq 0.8 \\ 0, & \psi < 0.8. \end{cases} \tag{2.158}$$

Either of the above functions is adequate to define approximately a suitable relationship between ψ and I.

Now, suppose we define the flux linkage corresponding to the air gap line as

$$\psi_{AD} = L_{AD} \cdot I_F. \tag{2.159}$$

But, from Figure 2.14

$$\psi_A = \psi_B - \psi_S = L_{AD} I_B - \psi_S. \tag{2.160}$$

The slope of the air gap line is given as

$$\tan\alpha = \frac{L_{AD}}{1} = \frac{\psi_s}{I_B - I_A} \tag{2.161}$$

or

$$I_B - I_A = \frac{\psi_s}{L_{AD}} \tag{2.162}$$

and

$$S_{GD} = \frac{I_B - I_A}{I_A} = \frac{\psi_S}{L_{AD} I_A} = \frac{\psi_S}{\psi_A}. \tag{2.163}$$

Then we may compute

$$\psi_A = \psi_B - \psi_s = L_{AD} I_B - S_{GD} \psi_A \tag{2.164}$$

or

$$\psi_A (1 + S_{GD}) = L_{AD} I_B \tag{2.165}$$

and finally

$$\psi_A = \frac{L_{AD}}{1+S_{GD}} I_B = L_{AD} I_A . \tag{2.166}$$

This relationship is shown graphically in Figure 2.16. Thus, as the machine saturates, we can replace the mutual inductance by a saturated mutual inductance

$$L_{AD} = \frac{L_{ADo}}{1+S_{GD}} \tag{2.167}$$

where

L_{AD} = the mutual inductance at any saturation
L_{ADo} = the mutual inductance at zero saturation.

A similar procedure could be followed for q-axis saturation, which would be different from the d-axis saturation because the mutual inductance is different in the q-axis. This procedure assumes that saturations in the two axes are independent, which is not strictly true but is a rough approximation to the truth, and is considered far better than ignoring the q-axis saturation. It is also true that saturation depends on the armature current components in the two axes and these currents affect the saturation directly.

2.13.1 Parameter Sensitivity to Saturation

In working with the machine equations we deal with many parameters that are defined in terms of the mutual inductance. Table 2.3 defines several d-axis parameters that are of interest and one could readily prepare a similar list of q-axis parameters. Examine Table 2.3 carefully. Suppose we assume that all leakage inductances are constants (although it has been shown that this is not strictly true, it is considered to be a very good approximation). Under this assumption, the only inductances in Figure 2.10 that saturate are the two mutual inductances. As the d-axis mutual inductance saturates, how does this affect the parameters in Table 2.3? To

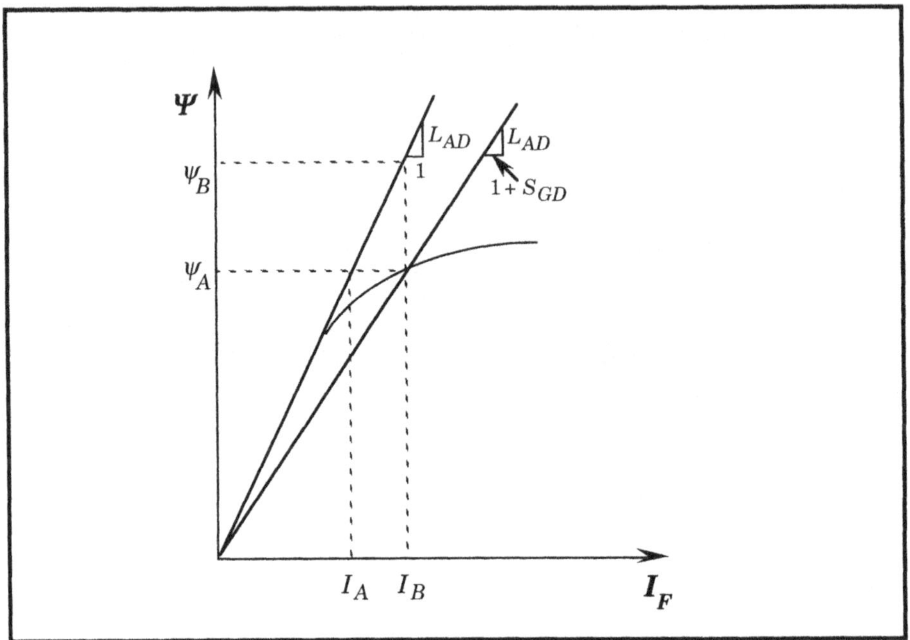

Figure 2.16 Computation of ψ_A by means of a "Saturated" Air Gap Line through the Operating Point

answer this question we compute the total differential of each parameter. In all cases we shall use the subscript "o" to indicate the unsaturated or zero current value of that parameter.

For example, examine the parameter

$$L_d = L_{AD} + \ell_a. \tag{2.168}$$

We adopt the notation for no saturation using "o" subscripts.

$$L_{do} = L_{ADo} + \ell_a. \tag{2.169}$$

Then

$$L_d = L_{do} + \Delta L_d \tag{2.170}$$

where

$$\Delta L_d = \left.\frac{\partial L_d}{\partial L_{AD}}\right|_o \Delta L_{AD} \tag{2.171}$$

but

$$\frac{\partial L_d}{\partial L_{AD}} = 1 \tag{2.172}$$

so that

$$L_d = L_{do} + \Delta L_{AD}. \tag{2.173}$$

Now, by definition

$$\Delta L_{AD} = L_{AD} - L_{ADo} \tag{2.174}$$

or

$$\Delta L_{AD} = \frac{L_{ADo}}{1+S_{DG}} - L_{ADo} = -\frac{S_{DG}L_{ADo}}{1+S_{DG}}. \tag{2.175}$$

Then

$$L_d = L_{do} - \frac{S_{DG}L_{ADo}}{1+S_{DG}}. \tag{2.176}$$

This process is easily repeated for the other parameters. The results are shown, for the d-axis quantities, in Table 2.6.

Note that some parameters change dramatically, and are actually amplified, by saturation. Most of these parameters change very little, however, for small changes in flux linkage. Similar conclusions may be drawn for the q-axis parameters. The important thing to be learned from the above is that these parameters are not "constants," but vary with the system operating state. In a transient condition, these parameters vary constantly and some of them should not be treated as constants at all.

2.13.2 Saturation in SSR Studies

In modeling the synchronous machine for SSR studies we concentrate on the small signal, linear performance. The open-circuit saturation curve

Table 2.6 Change in Machine Parameters with Saturation of L_{AD}

Parameter	Δ Parameter	Typical Sensitivity *
L_d	ΔL_{AD}	21%
L'_d	$\left(\frac{\ell_F}{L_{Fo}}\right)^2 \Delta L_{AD}$	1.5%
L''_d	$\left(\frac{L''_{do} - \ell_a}{L_{ADo}}\right)^2 \Delta L_{AD}$	0.3%
τ''_{do}	$\left(\frac{1}{\omega_B r_D} \frac{\partial L'_d}{\partial L_{AD}}\right) \Delta L_{AD}$	1.4%
τ'_{do}	$\left(\frac{1}{\omega_B r_F}\right) \Delta L_{AD}$	20%

* Change in parameter for 10% change in ψ_{AD}

shown in Figure 2.15 is a description of the total saturation of the machine. The saturation model, shown in Figure 2.16, should be regarded as a piecewise-linear model of the total saturation.

For small disturbances, the relationship between flux linkage and current is not along the straight line through the origin, as shown in Figure 2.16, but rather along a minor hysteresis loop as shown in Figure 2.17 [11].

Clearly, the slope of these minor hysteresis loops is much smaller than that of the major loop, which we formerly modeled approximately by the gap line. A comparison is provided by Minnich [12] and is reproduced in Table 2.7. These data indicate that the small signal value of the mutual inductance at rated voltage is only about 53% of the (unsaturated) large signal value. Stated another way, the slope of the minor hysteresis loop at rated voltage is about one-half the slope of the air gap line.

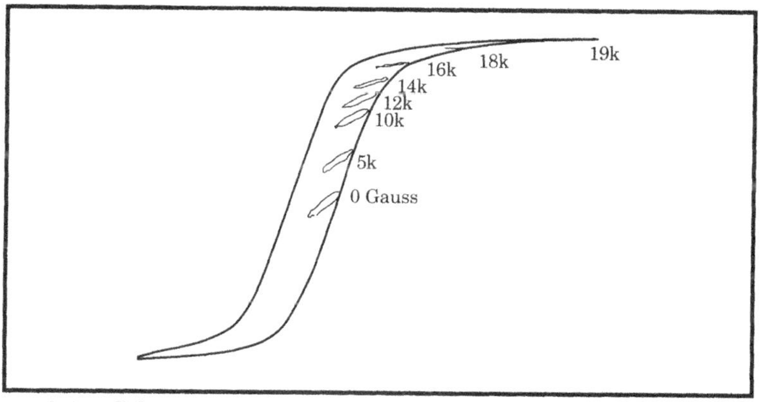

Figure 2.17 Schematic Incremental Minor Loops Superimposed on the Major Hysteresis Loop for Rotor Steel [12]

Table 2.7 Comparison of L_{AD} Small Signal and Large Signal Values at Open Circuit Conditions

Terminal Voltage per unit	L_{AD} in per unit	
	Small Signal	Large Signal
0	1.60	1.60
0.63	1.47	1.69
1.00	0.90	1.58

This is a remarkable result and could be very important in machine modeling for SSR studies. Minnich [12] makes the point quite clear in declaring

> In calculations which involve linearized small perturbations, such as eigenvalue calculations, it is perfectly legitimate to use the small-signal frequency response model -- small L_{AD} included. In fact it is rigorously correct to do so.

This does not mean that the total machine inductance, as seen from the network, will be greatly changed. The network sees the transient (or subtransient) inductance of the machine under transient (or subtransient) conditions and these inductances are very insensitive to saturation. This can be understood by examining the d- and q-axis equivalent circuits of Figure 2.10. The mutual inductance, which we consider to be the only element that saturates, is a large inductance and is in parallel with the much smaller leakage inductance. These leakage inductances carry most of the transient current and the exact value of the larger parallel mutual inductance is relatively unimportant, as far as the network coupling is concerned.

A practical procedure to use in eigenvalue studies is as follows:

1. Estimate the small signal L_{AD} to be 60% of the value normally given by the manufacturer for transient stability studies.

2. Compute the eigenvalues using both the small signal and large signal values of L_{AD} to make sure that this parameter is not critical to the solution process.

3. If the value of L_{AD} is shown to be critical, ask the machine manufacturer for an accurate small signal value. This can be determined by finite element techniques as described in [12].

2.14 REFERENCES FOR CHAPTER 2

1. IEC Std. 34-10-1975, "Rotating Electrical Machines. Part 10: Conventions for Description of Synchronous Machines," International Electrotechnical Commission, Geneva, 1975.

2. IEEE Std. 100-1984, **IEEE Standard Dictionary of Electrical and Electronics Terms**, IEEE, New York, 1984.

3. McPherson, George, **An Introduction to Electrical Machines and Transformers**, John Wiley and Sons, New York, 1981.

4. Anderson, P. M., and A. A. Fouad, **Power System Control and Stability**, Iowa State University Press, 1976.

5. Kimbark, Edward W., **Power System Stability, v 3, Synchronous Machines**, John Wiley and Sons, New York, 1956.

6. Park, R. H., "Two-reaction Theory of Synchronous Machines; Part I, Generalized Method Analysis," **Trans. AIEE**, v. 48, July 1929, p. 716-730.

7. Park, R. H., "Two-reaction Theory of Synchronous Machines; Part II," ibid, June 1933.

8. Hohn, F. E., **Elementary Matrix Algebra**, Macmillan, 1958.

9. Lewis, William A., **The Principles of Synchronous Machines**, 3rd Ed., Illinois Institute of Technology Bookstore, 1959.

10. Lewis, William A., "A Basic Analysis of Synchronous Machines," **Trans. AIEE**, Pt., 1, v. PAS-77, 1958, p. 436-455.

11. IEEE Committee Report, "Proposed Excitation System Definitions for Synchronous Machines," **Trans. IEEE**, v. PAS-88, 1969, p. 1248-1258.

12. Minnich, S. H., "Small Signals, Large Signals, and Saturation in Generator Modeling," **IEEE Trans. on Energy Conversion**, v. 86, March 1986, p. 94-102.

CHAPTER 3

THE NETWORK MODEL

The first experience of most engineers in writing circuit equations is to learn the methods of loop currents and node voltages, which are often learned as a set of rules that are easily remembered and that almost always work. In practice, these methods are applied to relatively simple circuits, and are not easily coded for digital computer analysis for the general case. As digital computers have become available to most engineers, there has been a trend away from these traditional methods of analysis, at least for large or complex circuits. Describing the network to the computer, however, requires a disciplined approach if one is to avoid the need to write a new computer program for each network to be analyzed [1]. One such approach began to emerge in the 1960's, in both network and systems theory, in the form of what has been termed the "state-space" approach.

The state-space approach to network analysis results in the differential equations that describe a linear network being arranged in the form

$$\begin{aligned} \dot{\mathbf{Y}}_1 &= \mathbf{A}\mathbf{Y}_1 + \mathbf{B}\mathbf{F} \\ \mathbf{Y}_2 &= \mathbf{C}\mathbf{Y}_1 + \mathbf{D}\mathbf{F} \end{aligned} \tag{3.1}$$

where

$$\begin{aligned} \mathbf{Y}_1 &= \text{the } n \text{ state variables} \\ \mathbf{F} &= \text{the } m \text{ input variables} \\ \mathbf{Y}_2 &= \text{the } r \text{ output variables} \\ \mathbf{A} &= \text{an } n \times n \text{ matrix} \\ \mathbf{B} &= \text{an } n \times m \text{ matrix} \\ \mathbf{C} &= \text{an } r \times n \text{ matrix} \\ \mathbf{D} &= \text{an } r \times m \text{ matrix.} \end{aligned} \tag{3.2}$$

These equations apply to lumped, linear, finite, time-invariant networks. The form of the equations is somewhat arbitrary in that it arranges the equations that describe the network as a set of n first order differential equations and r algebraic equations. It can be shown, however, that this form of equations is very convenient. There are several reasons that favor this form of network equations [2]:

1. It provides better insight into the physical aspects of the network problem than some other forms.
2. This form of equations has been thoroughly studied and there are known techniques for solving equations in this form, both analytically and by digital computer.
3. This form of equations is suitable for digital computer solution.
4. This form may be extended to nonlinear network equations in a straightforward manner.

Expressing the network equations in this standard form leads naturally to several important questions:

(a) What is the significance of the notion of the "state" in a network and what does this concept mean in terms of the physical network?

(b) How many states are required to completely describe the network, and how are these states to be defined?

(c) Is there a minimum number of states that are necessary and sufficient to describe the network and, if so, how is this minimum number to be found?

We will answer these questions in this chapter, where we present a brief review of the technique for writing the network state equations. The review material presented here is not intended to be a thorough development of the theory of state variable network descriptions. For this treatment, the reader is referred to the literature [1-4]. Our approach is to show, by example, how the state equations might be written, using an example that represents a small power system. Using this technique, the engineer should be able to grasp quickly the meaning of the network states, as well as the problems encountered in describing power systems by the state variable approach.

This chapter will not present material related to the *solution* of the state variable equations, but will simply point to references that describe techniques that may be used. As noted above, there has been a great deal of effort invested in the study of solution methods and in the preparation of computer programs to solve equations in the form of (3.1). Writing the network equations in this form permits us to take advantage of this growing body of work.

3.1 AN INTRODUCTORY EXAMPLE

The concept of a state-space approach to the analysis of networks will be aided by the examination of a simple power system network. In presenting a simple network and working through the process of writing the state equations, we will use those concepts from network theory that are required for a general formulation, and that might be used on any linear network of any size and of any complexity.

Consider the two machine power system shown in Figure 3.1. This network can be thought of as a generator (constant voltage source e_1) connected through a transmission line (R_2 and L_3 in series) to a constant impedance load (R_4 and L_5 in parallel), with this load also being served through another, series compensated, transmission line (R_6, L_7, and C_8 in series) by another generator (constant voltage source e_9). Note that every

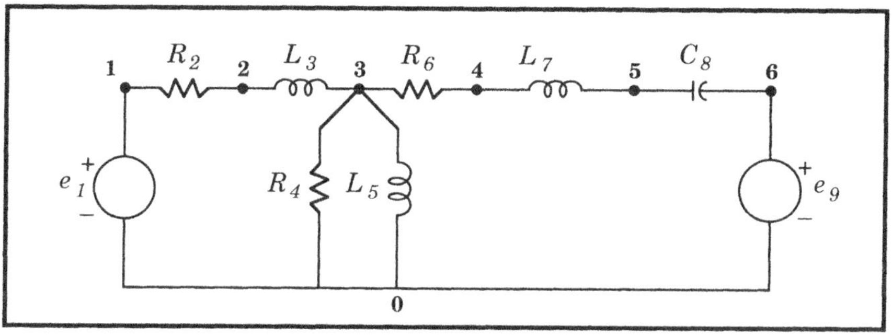

Figure 3.1 Example Network for State-Space Solution

element in the network is considered as a separate branch, and each of the branches and nodes are numbered, the reason for which will be apparent as the solution is developed. For this network, we have identified seven nodes, numbered from 0 through 6, and nine branches, numbered from 1 through 9. There are two independent voltage sources that have been arbitrarily assigned as branches 1 and 9 of the network. There are no independent current sources in this network.

The first task in solving the network is to draw an oriented or directed graph [1,2], which is shown in Figure 3.2. The arrow associated with each branch is the assumed direction of current flow in that branch as well as the direction of the assumed voltage drop across that branch. Note that both the branches and the nodes are numbered in an arbitrary sequence.

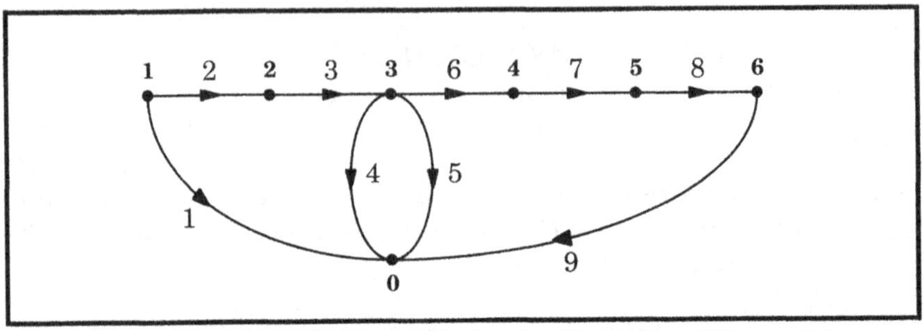

Figure 3.2 A Directed Graph for the Network of Figure 3.1

Having constructed an oriented graph, it is possible to define the network topology in terms of the incidence matrix, which tells how the network is connected. The method for doing this is illustrated in many textbooks on network theory [1-4]. We will not pursue this explicitly here, as it is assumed that the reader is acquainted with the method. This step is not absolutely essential for reaching our goal for writing the equations in the desired form, but it is essential background theory.

The next task is to construct a *proper tree* of the directed graph, where a proper tree is defined as follows:

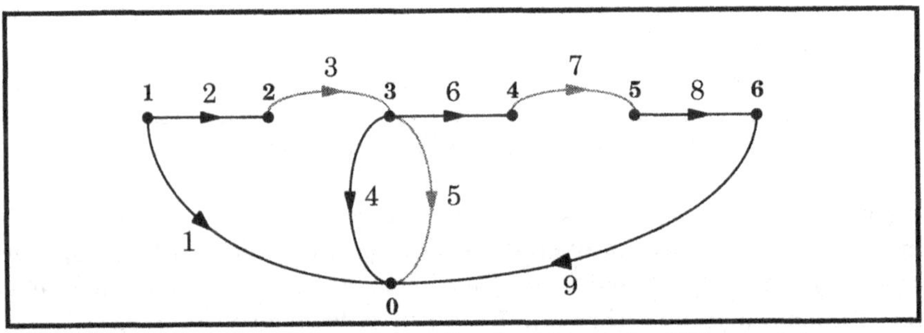

Figure 3.3 A Proper Tree for the Network of Figure 3.1

> A proper tree for a connected lumped network, composed of linear, time-invariant, resistance, inductance, and capacitance non-source elements and independent voltage and current sources, is a tree that contains all voltage sources as tree branches, all current sources as link branches, and as many capacitor tree branches and inductive line branches as possible [3].

For the ideal case, this definition allows resistance branches among the tree branches, but no inductance branches or independent current sources as tree branches. The link branches, then, will contain all the inductance branches, all independent current sources, and usually some of the resistance branches. A proper tree for the network of Figure 3.1 is shown in Figure 3.3. The tree branches are those branches shown in solid black and the link branches are shown in gray.

Note that this network contains two independent voltage sources and one capacitor branch, and that these elements are all tree branches in the proper tree. The remaining tree branches are the three resistor branches. The link branches are the three inductances. This network *can* be described by a proper tree. We shall see later that not all networks can be described by a proper tree, which leads to certain complexities in the equation formulation.

Given a proper tree, the state-space equations will always have the form

$$\begin{bmatrix} p\mathbf{i}_L(t) \\ p\mathbf{v}_C(t) \end{bmatrix} = \mathbf{A}\begin{bmatrix} \mathbf{i}_L(t) \\ \mathbf{v}_C(t) \end{bmatrix} + \mathbf{B}\begin{bmatrix} \mathbf{i}_I(t) \\ \mathbf{v}_V(t) \end{bmatrix} \tag{3.3}$$

where

$\mathbf{i}_L$ = The vector of all inductance (link) currents

$\mathbf{v}_C$ = The vector of all capacitance (tree) voltages

$\mathbf{i}_I$ = The vector of all independent current sources

$\mathbf{v}_V$ = The vector of all independent voltage sources

and where p is the derivative operator. Note that the network state variables are the capacitor voltages and the inductor currents.

For the network of Figure 3.1, we may write the following constraints from Kirchhoff's voltage law (KVL) and current law (KCL), where the voltages are voltage drops across the branches in the positive current direction and the currents are the defined branch currents in the direction defined in the directed graph.

KVL:

Write the voltage for link branch k in terms of tree branch voltages:

$$v_{LBk} = \sum v_{TBk}. \tag{3.4}$$

For the network of Figure 3.3:

$$
\begin{aligned}
v_3 &= -v_2 + v_1 - v_4 \\
v_5 &= v_4 \\
v_7 &= -v_6 + v_4 - v_9 - v_8
\end{aligned}
\tag{3.5}
$$

Note that these equations express the voltages of the link branches, on the left side of (3.5), in terms of the tree branch voltages, on the right side.

KCL:
Write the tree branch currents for branch j in terms of link currents:

$$
i_{TBj} = \sum i_{LBj}. \tag{3.6}
$$

For the network of Figure 3.3:

$$
\begin{aligned}
i_1 &= -i_3 \\
i_2 &= i_3 \\
i_4 &= i_3 - i_5 - i_7 \\
i_6 &= i_7 \\
i_8 &= i_7 \\
i_9 &= i_7.
\end{aligned}
\tag{3.7}
$$

The currents on the left are tree branch currents, each of which is defined in terms of the link branch currents.

In addition to the above constraints that describe the network topology, we may write the physical voltage-current equations for each branch, which we refer to as the v-i equations.

***v-i* Equations:**

$$
\begin{aligned}
v_1 &= e_1 & & \\
v_2 &= R_2 i_2 & v_6 &= R_6 i_6 \\
v_3 &= L_3 \frac{di_3}{dt} & v_7 &= L_7 \frac{di_7}{dt} \\
v_4 &= R_4 i_4 & i_8 &= C_8 \frac{dv_8}{dt} \\
v_5 &= L_5 \frac{di_5}{dt} & v_9 &= e_9.
\end{aligned}
\tag{3.8}
$$

There is one equation for each branch. Note that the only equations that express derivatives are those that involve energy storage elements, the inductances and capacitances. *We define these variables to be the state variables.* To get the equations in the desired standard form, shown in (3.1), we manipulate the equations as follows.

From the *v-i* equations, we write each derivative and replace the right hand side with variables substituted from the KVL and KCL equations, and from the remaining *v-i* equations. Performing this operation, we may write the following equations.

$$\begin{aligned}
L_3\frac{di_3}{dt} &= v_3 = v_1 - v_2 - v_4 = e_1 - R_2 i_3 - R_4(i_3 - i_5 - i_7)\\
L_5\frac{di_5}{dt} &= v_5 = v_4 = R_4(i_3 - i_5 - i_7)\\
L_7\frac{di_7}{dt} &= v_7 = v_4 - v_6 - v_8 - e_9\\
&= R_4(i_3 - i_5 - i_7) - R_6 i_7 - v_8 - e_9\\
C_8\frac{dv_8}{dt} &= i_8 = i_7.
\end{aligned} \tag{3.9}$$

The right hand side of each equation now contains only the state variables, the inductor currents and capacitor voltages, and the input functions, the independent generator voltages.

Rearranging (3.9) in matrix form, we write

$$\begin{bmatrix} L_3 & & & \\ & L_5 & & \\ & & L_7 & \\ & & & C_8 \end{bmatrix}\begin{bmatrix} pi_3 \\ pi_5 \\ pi_7 \\ pv_8 \end{bmatrix} = \begin{bmatrix} -(R_2+R_4) & +R_4 & +R_4 & 0 \\ +R_4 & -R_4 & -R_4 & 0 \\ +R_4 & -R_4 & -(R_4+R_6) & -1 \\ 0 & 0 & +1 & 0 \end{bmatrix}\begin{bmatrix} i_3 \\ i_5 \\ i_7 \\ v_8 \end{bmatrix} + \begin{bmatrix} 1 & 0 \\ 0 & 0 \\ 0 & -1 \\ 0 & 0 \end{bmatrix}\begin{bmatrix} e_1 \\ e_9 \end{bmatrix}. \tag{3.10}$$

The coefficient matrices are made up of constants that depend only on the values of the network resistors, capacitors, and inductors. The matrix on the left of the derivative vector is diagonal and therefore has an inverse that may be found by inspection. Premultiplying by the inverse of this matrix, we compute the final form of the state equations as

$$\begin{bmatrix} pi_3 \\ pi_5 \\ pi_7 \\ pv_8 \end{bmatrix} = \begin{bmatrix} \frac{-(R_2+R_4)}{L_3} & \frac{+R_4}{L_3} & \frac{+R_4}{L_3} & 0 \\ \frac{+R_4}{L_5} & \frac{-R_4}{L_5} & \frac{-R_4}{L_5} & 0 \\ \frac{+R_4}{L_7} & \frac{-R_4}{L_7} & \frac{-(R_4+R_6)}{L_7} & \frac{-1}{L_7} \\ 0 & 0 & \frac{+1}{C_8} & 0 \end{bmatrix} \begin{bmatrix} i_3 \\ i_5 \\ i_7 \\ v_8 \end{bmatrix} + \begin{bmatrix} \frac{1}{L_3} & 0 \\ 0 & 0 \\ 0 & \frac{-1}{L_7} \\ 0 & 0 \end{bmatrix} \begin{bmatrix} e_1 \\ e_9 \end{bmatrix} \tag{3.11}$$

or

$$\dot{\mathbf{Y}}_1 = \mathbf{A}\mathbf{Y}_1 + \mathbf{B}\mathbf{F}. \tag{3.12}$$

No output equation has been written because the output variables of interest have not been defined. Suppose we define the required output variable to be the voltage across the passive load at bus 3. This voltage may be written in several ways, the simplest of which is the following.

$$y_2 = \begin{bmatrix} +R_4 & -R_4 & -R_4 & 0 \end{bmatrix} \begin{bmatrix} i_3 \\ i_5 \\ i_7 \\ v_8 \end{bmatrix} \tag{3.13}$$

or

$$\mathbf{Y}_2 = \mathbf{C}\mathbf{Y}_1 + \mathbf{D}\mathbf{F}. \tag{3.14}$$

In this case, the output equation is scalar since only one output variable is defined. Also, we note that this output variable is a function of the state variables only and is not a function of the input variables, i.e., the matrix **D** is the null matrix.

We are now able to answer some of the questions posed earlier.

(a) The network states are the inductor currents and the capacitor voltages, which govern the rate at which energy is stored in the network.

(b) The network under study has four energy storage elements, three inductors, and one capacitor. Hence, it requires four state variables in order to completely define the network. It should be pointed out again that this is a special case, since this network can be described in terms of a proper tree. A proper tree cannot be found for every network, hence this conclusion is not always true.

(c) The question regarding the minimum number of states is not obvious from this simple example. More will be said on this later.

We may state the above observations more formally as follows. The network equations have been written in the form of the KVL, KCL, and v-i equations of the network branches. In a formal sense, the KVL and KCL equations can be stated in terms of the fundamental cutset matrix and the fundamental loop matrix equations of the network.

First, for the voltage constraints, we write the following matrix equations [4] that express the Kirchhoff fundamental voltage law loop relationships.

$$\mathbf{B}_f \mathbf{v}_b = \mathbf{0} \tag{3.15}$$

where

$\mathbf{B}_f =$ The fundamental loop matrix

$\mathbf{v}_b =$ The branch voltage vector.

Also, the Kirchhoff's fundamental current law equations may be written as

$$\mathbf{Q}_f \mathbf{i}_b = \mathbf{0} \tag{3.16}$$

where

$\mathbf{Q}_f =$ The fundamental cut set matrix

$\mathbf{i}_b =$ The branch currents.

Both (3.15) and (3.16) have $b = 9$ unknowns for a total of $2b$ unknowns. Equation (3.15) provides three independent equations in these nine unknowns, equal to the number of link branches. Equation (3.16) gives another six equations, equal to the number of tree branches. Together, then, these equations provide $b = 9$ equations in $2b$ unknowns. The remaining b equations come from the branch v-i equations, with one equation for each of the b branches, thus giving exactly $2b$ equations in $2b$ unknowns.

The fundamental branch matrices can both be computed from a network incidence matrix [1-4]. For the simple problem presented here, we can show by inspection of the proper tree that these matrices are given by the following equations.

$$
\mathbf{Q}_f = \begin{array}{c} \\ f1 \\ f2 \\ f4 \\ f6 \\ f8 \\ f9 \end{array}
\begin{array}{c}
\begin{array}{ccc:cccccc} i3 & i5 & i7 & i1 & i2 & i4 & i6 & i8 & i9 \end{array} \\
\left[\begin{array}{ccc:cccccc}
+1 & & & 1 & & & & & \\
-1 & & & & 1 & & & & \\
-1 & +1 & +1 & & & 1 & & & \\
 & & -1 & & & & 1 & & \\
 & & -1 & & & & & 1 & \\
 & & -1 & & & & & & 1
\end{array}\right]
\end{array}
\tag{3.17}
$$

$$
\mathbf{B}_f = \begin{array}{c} \\ c3 \\ c5 \\ c7 \end{array}
\begin{array}{c}
\begin{array}{ccc:cccccc} v3 & v5 & v7 & v1 & v2 & v4 & v6 & v8 & v9 \end{array} \\
\left[\begin{array}{ccc:cccccc}
1 & & & -1 & +1 & +1 & & & \\
 & 1 & & & & -1 & & & \\
 & & 1 & & & -1 & +1 & +1 & +1
\end{array}\right]
\end{array}
\tag{3.18}
$$

Note that each of these matrices contains an identity matrix and that the remainder is either 6 x 3 or 3 x 6, with these submatrices being the negative transpose of each other.

The important thing to remember here is that the network solution follows from a straightforward application of topological and branch equations, which are well suited for computer manipulation. It should be emphasized that this network is a special one, since it *can* be described by a *proper tree*. We now examine a case that violates the rules for a proper tree.

3.2 THE DEGENERATE NETWORK

A network is said to be *degenerate* if one or both of the following conditions are satisfied [4]:

1. It contains a circuit (loop) composed only of capacitors and/or independent or dependent voltage sources.

2. It contains a cutset composed only of inductors and/or independent or dependent current sources.

Networks that are degenerate cannot be described by a proper tree. We illustrate this by an extension of the example used in Section 3.1.

The small power system shown in Figure 3.1 is modified in two ways, as follows:

(a) A shunt capacitor bank is connected to node 5.

(b) The load at node 3 is changed to a shunt reactance by removing the shunt resistance element, R_4.

The resulting network is shown in Figure 3.4, where it is noted that branch 4 is now missing and a new branch, branch 10, has been added.

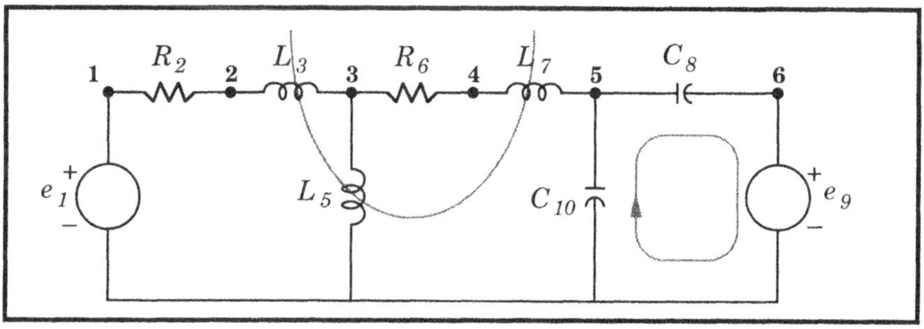

Figure 3.4 The Modified System Network

An examination of the network shown in Figure 3.4 reveals that it violates two of the rules for constructing a proper tree:

1. There is a loop containing nothing but capacitors and a voltage source.

2. There is a cutset of only link branch inductors.

These violations are highlighted in Figure 3.4. Therefore, we might expect that the simple rule for selecting the state variables as the inductor currents and capacitor voltages will not work. On closer examination, this should be obvious. The capacitor voltages are no longer independent because of the loop of capacitors and the voltage source that allows one capacitor voltage to be expressed in terms of the other. The same can be said of the inductor currents, since only two of the three can possibly be independent currents, since the three currents add to zero at node 3.

To analyze the network, we again choose a tree, following the rules for a proper tree as closely as possible. The tree selected is shown in Figure 3.5.

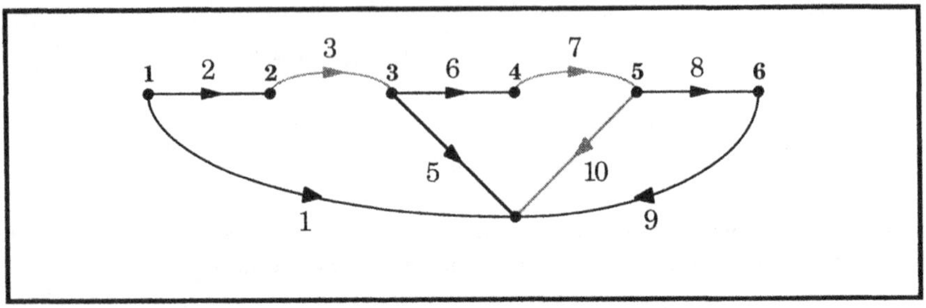

Figure 3.5 A Tree for the Circuit of Figure 3.4

The tree in Figure 3.5 violates the rules for a proper tree in two ways. First, the inductor, branch 5, in now a tree branch. On closer examination, one of the inductors *must* be used as a tree branch. Also, the capacitor of branch 10 is a link branch. Indeed, one of the capacitors *must* be declared to be a link branch. We proceed as before to write the network equations.

KVL:

$$
\begin{aligned}
v_3 &= -v_2 + v_1 - v_5 \\
v_7 &= -v_6 + v_5 - v_{10} \\
v_{10} &= v_8 + v_9
\end{aligned}
\tag{3.19}
$$

KCL:

$$
\begin{aligned}
i_1 &= -i_3 \\
i_2 &= i_3 \\
i_5 &= i_3 - i_7 \\
i_6 &= i_7 \\
i_8 &= i_7 - i_{10} \\
i_9 &= i_8 = i_7 - i_{10}
\end{aligned}
\tag{3.20}
$$

***v-i* equations:**

$$
\begin{aligned}
v_1 &= e_1 & v_7 &= L_7 \frac{di_7}{dt} \\
v_2 &= R_2 i_2 & i_8 &= C_8 \frac{dv_8}{dt} \\
v_3 &= L_3 \frac{di_3}{dt} & v_9 &= e_9 \\
v_5 &= L_5 \frac{di_5}{dt} & i_{10} &= C_{10} \frac{dv_{10}}{dt}. \\
v_6 &= R_6 i_6
\end{aligned}
\tag{3.21}
$$

The independent state variables are now selected to be the tree capacitor voltage and the link inductors currents, exactly as in the previous case. This gives the following equations:

$$\begin{aligned}
L_3 \frac{di_3}{dt} &= v_3 = v_1 - v_2 - v_5 = e_1 - R_2 i_2 - L_5 \frac{di_5}{dt} \\
&= e_1 - R_2 i_3 - L_5 \frac{d(i_3 - i_7)}{dt} \\
L_7 \frac{di_7}{dt} &= v_7 = v_5 - v_6 - v_{10} \\
&= L_5 \frac{d(i_3 - i_7)}{dt} - R_6 i_7 - v_8 - e_9 \\
C_8 \frac{dv_8}{dt} &= i_8 = i_7 - i_{10} = i_7 - C_{10} \frac{d(v_8 + v_9)}{dt} \\
&= i_7 - C_{10} \frac{dv_8}{dt} - C_{10} \frac{de_9}{dt}
\end{aligned} \tag{3.22}$$

Rearranging, we write (3.22) in matrix form as follows.

$$\begin{bmatrix} (L_3 + L_5) & -L_5 & 0 \\ -L_5 & (L_5 + L_7) & 0 \\ 0 & 0 & (C_8 + C_{10}) \end{bmatrix} \begin{bmatrix} pi_3 \\ pi_7 \\ pv_8 \end{bmatrix} = \begin{bmatrix} -R_2 & 0 & 0 \\ 0 & -R_6 & -1 \\ 0 & +1 & 0 \end{bmatrix} \begin{bmatrix} i_3 \\ i_7 \\ v_8 \end{bmatrix} + \begin{bmatrix} 1 & 0 \\ 0 & -1 \\ 0 & 0 \end{bmatrix} \begin{bmatrix} e_1 \\ e_9 \end{bmatrix} + \begin{bmatrix} 0 & 0 \\ 0 & 0 \\ 0 & -C_{10} \end{bmatrix} \begin{bmatrix} pe_1 \\ pe_9 \end{bmatrix}. \tag{3.23}$$

The coefficient matrix on the left clearly has an inverse. Multiplying both sides of (3.23) by this inverse, we obtain the state-space equation of the network in the form

$$\begin{aligned}
\dot{\mathbf{Y}}_1 &= \mathbf{A}\mathbf{Y}_1 + \mathbf{B}\mathbf{F} + \mathbf{B}_1\dot{\mathbf{F}} \\
\mathbf{Y}_2 &= \mathbf{C}\mathbf{Y}_1 + \mathbf{D}\mathbf{F} + \mathbf{D}_1\dot{\mathbf{F}}
\end{aligned} \tag{3.24}$$

where the new equation form includes terms involving the derivative of the inputs. This is often the case in power systems, where inductance cutsets are very likely to occur and where capacitance loops are possible for circuits that contain series capacitors. In other words, the example network is not at all unusual for a power system network.

We note from (3.23) that the number of state variables is only three, but the number of storage elements is five. This is compared to the circuit of Figure 3.1, where the number of independent states is four, which is exactly equal to the number of energy storage elements. How are we to know that (3.23) represents the absolute minimum number of states and that one of the state variables chosen is not, in fact, dependent on some linear combination of the other state variables? The answer to this question comes from network theory, which will be reviewed in the next section.

State equations of the form (3.24) can be converted to the standard or normal form (3.1) by the transformation [4]

$$\begin{aligned} \mathbf{Y}_{1new} &= \mathbf{Y}_1 - \mathbf{B}_1\mathbf{F} \\ \mathbf{Y}_{2new} &= \mathbf{Y}_2 - \mathbf{D}_1\dot{\mathbf{F}} \end{aligned} \tag{3.25}$$

which, when substituted into (3.24) gives a new equation in the new state variables that is exactly in the form of (3.1). This means that the defined states are no longer the capacitor voltages and inductor currents, but are defined by (3.25) as some linear combination of these variables. This is perfectly all right, since the independent states may be selected in many different ways. If it is desired to retain the capacitor voltages and inductor currents as state variables, there is no reason why the form (3.24) can't be retained. The only problem is to eliminate those capacitor voltages and inductor currents that are dependent in writing (3.24).

3.3 THE ORDER OF COMPLEXITY OF THE NETWORK

The order of complexity of a network is defined as the number of state variables, or the number of independent, coupled, first-order linear differential equations, that are required for the *reduced representation* of the network [2]. Here, the *reduced representation* means the number of states that remain after eliminating all state variables that are dependent on other states, as shown in the preceding section. This number is also equal to the number of independent initial conditions that can be specified, and is also equal to the number of natural frequencies of the network, counting each frequency according to its multiplicity [4].

The number of independent states can be represented by the following formula.

$$n = n_L + n_C - m_L - m_C \tag{3.26}$$

where

n = number of independent states

n_L = number of inductors

n_C = number of capacitors

m_L = number of independent all– inductive cutsets

m_C = number of independent all– capacitive loops.

Note that the number m_C of capacitive loops must be independent, that is, each new such loop counted must contain at least one new capacitor. A similar statement can be made for the inductive cutsets. These numbers may be obtained by inspection in a small network. The number m_L is found by shorting all resistors, capacitors, and voltage sources and noting the remaining currents through inductors. The number m_C can be found by opening all resistors, inductors, and current sources and noting the remaining loops of capacitors and voltage sources. Computer programs have been prepared that perform this function in a rigorous manner [3].

If we apply equation (3.26) to the network of Figure 3.4, we have the following expression:

$$\begin{aligned} n &= n_L + n_C - m_L - m_C \\ &= 3 + 2 - 1 - 1 = 3 \end{aligned} \tag{3.27}$$

which is exactly the number of independent equations that we found by inspection of the network, namely, three. Note also that the circuit of Figure 3.1 gives four independent states, which is equal to the number of inductors plus the number of capacitors.

Finally, we note that the order of complexity of a network has the same meaning as the *order of a system* in linear system or control theory, which refers to the number of independent differential equations required to uniquely define the system. In mechanical systems, this same concept is often termed the *degrees of freedom*, which refers to the number of independent position variables that can be defined in a mechanical system. These terms have exactly the same meaning, and all refer to the rank n of the matrix $\mathbf{A}$ in the state variable equation. For a network, this order may be determined by knowing only the topology of the *R*-*L*-*C* branches and the independent sources [3,4].

It has been shown that this same method of analysis can be extended to circuits that contain transformers and dependent sources. This is important in power systems, where we often represent a transformer by a tee equivalent, thereby creating an inductor cutset. The representation of dependent sources in a power system will permit the analysis of circuits such as the *d* and *q* equivalent circuits of a synchronous machine, as shown in Figure 2.9. These circuits contain current controlled voltage sources, and they can be analyzed by the state-space network techniques described above [1,3,5].

3.4 FINDING THE NETWORK STATE EQUATIONS

This section presents the differential equations for electric transmission networks and develops the *d*-*q* transformations for these equations. The approach used is one that is convenient for the user of a computer program in SSR analysis, and is somewhat different from the matrix approach used in large-scale network solutions.

The transformation of network equations in the *a*-*b*-*c* reference frame to the 0-*d*-*q* reference frame presents a problem that must be recognized at the outset. The Park transformation, given by equation 2.48, includes the variable θ, the angular position of the generator shaft, which is defined by the relationship [5].

$$\theta = \omega_B t + \delta + \frac{\pi}{2} \quad \text{electrical radians} \tag{3.28}$$

where

ω_B = Base radian frequency, rad / s

$= 2\pi f_B = 120\pi$ for 60 Hz systems

t = Time in s

δ = Angle of deviation from a synchronously rotating reference with respect to the q axis, in electrical radians.

The derivative of (3.28) is given by

$$\omega = \frac{d\theta}{dt} = \omega_B + \frac{d\delta}{dt} \quad \text{electrical radians} \tag{3.29}$$

which is clearly the generator shaft angular velocity. This angular velocity enters the Park's transformation through the speed voltage term, as shown in (2.63). As noted in Section 2.11 (Linearization) the derivative $d\delta/dt$ is usually quite small compared to the base radian frequency. For a stable machine, the angular velocity ω oscillates about the base angular velocity following a disturbance. Clearly, the Park's transformation of the a-b-c reference frame equations introduces a term that is proportional to this angular velocity.

In applying Park's transformation to the network equations, however, we face a dilemma. Except for the rather trivial one-machine problem, there is no single angular velocity but a unique ω for each generator. If one generator is predominantly large, it would make sense to use the ω of that machine in the Park's transformation of the network equations. Usually, however, there are many machines of about the same size and no one machine dominates the system frequency. Indeed, all machines oscillate about the system base frequency, for a stable condition.

We resolve this problem by using the *base* angular velocity in the Park's transformation of the network equations. For a network of finite size, and with generation matched to the load, this base angular velocity is the weighted average of the angular velocities of all generators (see [9], Chapter 3).

The general problem is to find the minimum set of differential equations for a network of interconnected resistors, inductors, capacitors, and current sources of the following form.

$$\begin{aligned} \dot{\mathbf{Y}}_1 &= \mathbf{A}\mathbf{Y}_1 + \mathbf{B}\mathbf{F} + \mathbf{B}_1\dot{\mathbf{F}} \\ \mathbf{Y}_2 &= \mathbf{C}\mathbf{Y}_1 + \mathbf{D}\mathbf{F} + \mathbf{D}_1\dot{\mathbf{F}}. \end{aligned} \tag{3.30}$$

The algorithm used here will work with a much more general network than the one needed to represent the electric transmission network of a power system. For the power system case, we need only the three passive elements, resistors, inductors, and capacitors, plus current sources to represent the generators. The inputs to the network model, the $\mathbf{F}$ variables and their derivatives in (3.30), are the injection currents and their derivatives from the model for generators derived in Chapter 2. As long as the network is restricted to these types of elements, no derivatives higher than the first will be needed. The output of the network model, the $\mathbf{Y}_2$ variables in (3.30), are the voltages at the generator terminals. These serve as input variables to the generator model, completing the coupling between those two elements of the system.

The algorithm used here for obtaining the state equations was developed by Pottle [3]. It is a modification of that proposed by Dervisoglu [6], which is described in detail in several places, for example, in [5] and [7]. Pottle has developed a program based on this algorithm, which has been named CORNAP (Cornell Network Analysis Program), and which is described in [8]. Only that part of the program that forms the state equations is used directly in the work described here. It is a rather complex process and will not be described in detail, but an outline, quoted from [8], follows:

> We will assume that the state equations sought have a subset of the capacitor voltages and inductor currents as state variables. The problem is, of course, to proceed from an easily given description of the network to a set of first order coupled differential equations which contain only state variables and independent source inputs. All other branch voltages and currents are removed along the way. A summary of the method follows.
>
> 1. Form the fundamental loop matrix with respect to a normal tree[1], that is, a tree which contains all voltage sources as tree branches, all current sources as links, and as many capacitive tree branches and inductive links as possible. This matrix expresses a relation giving all tree-branch currents in terms of link currents and, using its negative transpose, giving all link voltage in terms of tree-branch voltages. The topological concepts necessary as background for the above statement are not negligible, but are nowhere near the level often encountered in introductory graduate courses.
>
> 2. Form and solve a set of algebraic equations relating
>   a) resistive tree-branch voltages to their currents (and henceto link currents),
>   b) resistive link currents to their voltages (and hence to tree-branch voltages),
>   c) controlled sources to their controlling quantities (and hence to link currents and tree-branch voltages).

[1]A normal tree and a proper tree are the same. Pottle uses the word "normal" in the reference quoted.

3. Form a set of initial state equations obtained by replacing inductor voltages by expressions involving derivatives of inductor currents and replacing capacitive currents by expressions involving derivatives of capacitor voltages.

4. By row reduction techniques, obtain the final state equations with dependent state variables removed.

The formulation of the network equations is illustrated by a very simple example.

Example 3.1 Consider the three-phase system shown in Figure 3.6 where a generator supplies a passive load through a single series-compensated transmission line. This is the simplest of all circuits to analyze, but gives a valuable overview of the detailed equations.

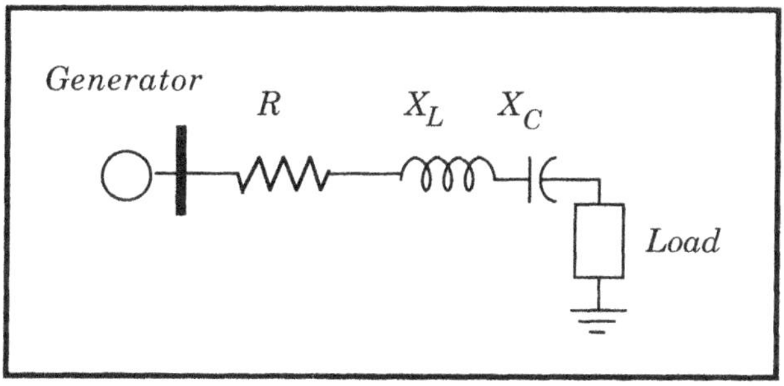

Figure 3.6 The Network For Example 3.1

The load is passive and is assumed to consist of inductance and resistance elements. The load, therefore, may be modeled as a series resistance and inductive reactance, which is simply added to the line components of the same type.

Figure 3.7 shows the resulting network for phase a only. Similar circuits could be drawn for phases b and c. Note that the three phases are assumed to be uncoupled, that is, there are assumed to be no mutual inductances or capacitances between phases. This makes the voltage drops in each phase a function of only the current in that phase. This clearly is a simplifying assumption that is not correct in physical transmission lines, but is used here to simplify not just the phase equations but the 0-d-q transformation as well.

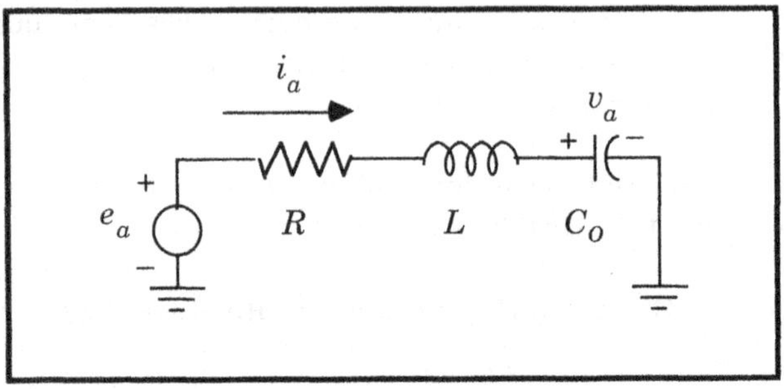

Figure 3.7 Circuit Diagram for Phase a

In Figure 3.7 we define the current as leaving the generator and the capacitor voltage v_a with the polarity shown. For the three-phase system we write for phase a

$$
\begin{aligned}
e_a &= Ri_a + L\frac{di_a}{dt} + v_a \\
i_a &= C_o\frac{dv_a}{dt}
\end{aligned}
\tag{3.31}
$$

and similarly for the other two phases. In matrix notation, we write all three equations as

$$
\begin{aligned}
\mathbf{e}_{abc} &= \mathbf{R}\mathbf{i}_{abc} + \mathbf{L}p\mathbf{i}_{abc} + \mathbf{v}_{abc} \\
\mathbf{i}_{abc} &= \mathbf{C}_o p\mathbf{v}_{abc}.
\end{aligned}
\tag{3.32}
$$

These equations are exactly in the form (3.30) with the following interpretations of the variables. First, we define the state variables and the input variables as

$$
\begin{aligned}
\mathbf{Y}_1 &= \mathbf{v}_{abc} \\
\mathbf{F} &= \mathbf{i}_{abc} \\
\dot{\mathbf{F}} &= \frac{d\mathbf{i}_{abc}}{dt}.
\end{aligned}
\tag{3.33}
$$

Next, we define the matrices as for the first of equations (3.30) as

$$
\begin{aligned}
&\mathbf{A} = \mathbf{0} \\
&\mathbf{B} = \mathbf{C}_o^{-1} \\
&\mathbf{B}_1 = \mathbf{0}
\end{aligned} \qquad (3.34)
$$

For the second of equations (3.30) we have

$$
\begin{aligned}
&\mathbf{C} = \mathbf{U} \\
&\mathbf{D} = \mathbf{R} \\
&\mathbf{D}_1 = \mathbf{L}.
\end{aligned} \qquad (3.35)
$$

This simple example is interesting since it points out clearly that the currents through the inductances and the voltages across the capacitors cannot *both* be state variables. We know that this is a third order system when all three phases are considered. The state variables are taken here to be the capacitor voltages, as noted in (3.33). The inductance currents are the same as the generator currents, which are considered as network input variables in the three-phase state-space formulation. The generator voltages are the network "output" variables $\mathbf{e_{abc}}$, which are the input variables to the machine equations. This completes the Example 3.1.

3.5 TRANSFORMING THE STATE EQUATIONS

In (3.30) of the previous section, the variables were considered in general with no specification of the reference frame. Now assume that equation (3.30) is written in terms of the *a-b-c* frame of reference. In this case the equation may be written as follows:

$$
\begin{aligned}
\dot{\mathbf{Y}}_{1abc} &= \mathbf{A}\mathbf{Y}_{1abc} + \mathbf{B}\mathbf{F}_{abc} + \mathbf{B}_1\dot{\mathbf{F}}_{abc} \\
\mathbf{Y}_{2abc} &= \mathbf{C}\mathbf{Y}_{1abc} + \mathbf{D}\mathbf{F}_{abc} + \mathbf{D}_1\dot{\mathbf{F}}_{abc}.
\end{aligned} \qquad (3.36)
$$

These equations can be transformed to the *0-d-q* reference frame in exactly the same way as the voltage equations for the generator model in Section 2.4. Multiplying the first equation by $\mathbf{P}$ and using equations similar to (2.59) for the transformation of the derivative terms gives the desired transformation. To simplify the computer algorithm, this is broken into three parts.

First, consider the derivative of the state variables. If we define the right side of the first equation to be $\mathbf{f_{abc}}$ we write

$$\dot{\mathbf{Y}}_{1abc} = \mathbf{f}_{abc}. \tag{3.37}$$

Transforming, we write

$$\mathbf{P}\dot{\mathbf{Y}}_{1abc} = \mathbf{P}\mathbf{f}_{abc} = \mathbf{f}_{0dq}. \tag{3.38}$$

We may write the derivative of the transformation as

$$\dot{\mathbf{Y}}_{10dq} = \frac{d}{dt}\mathbf{P}\mathbf{Y}_{1abc} = \mathbf{P}\dot{\mathbf{Y}}_{1abc} + \dot{\mathbf{P}}\mathbf{Y}_{1abc}. \tag{3.39}$$

From this we write the **P** transform of the derivative as

$$\mathbf{P}\dot{\mathbf{Y}}_{1abc} = \dot{\mathbf{Y}}_{10dq} - \dot{\mathbf{P}}\mathbf{Y}_{1abc} = \dot{\mathbf{Y}}_{10dq} - \dot{\mathbf{P}}\mathbf{P}^{-1}\mathbf{Y}_{10dq}. \tag{3.40}$$

Therefore, the **P** transform of (3.37), from (3.38) and (3.40) is given by

$$\mathbf{f}_{0dq} = \dot{\mathbf{Y}}_{10dq} - \dot{\mathbf{P}}\mathbf{P}^{-1}\mathbf{Y}_{10dq}. \tag{3.41}$$

Now, recall that

$$\dot{\mathbf{P}}\mathbf{P}^{-1} = \begin{bmatrix} 0 & 0 & 0 \\ 0 & 0 & -\omega \\ 0 & +\omega & 0 \end{bmatrix}. \tag{3.42}$$

This may be substituted into (3.41) with the result

$$\mathbf{f}_{0dq} = \begin{bmatrix} p & 0 & 0 \\ 0 & p & +\omega \\ 0 & -\omega & p \end{bmatrix} \mathbf{Y}_{10dq} = \mathbf{H}\mathbf{Y}_{10dq} \tag{3.43}$$

where we define the operational matrix **H** which forms the coupling between the d- and q-axis equations.

Usually, we are interested only in balanced systems in which the zero sequence may be neglected. For this case, we write (3.43) as

$$\mathbf{f}_{dq} = \begin{bmatrix} p & +\omega \\ -\omega & p \end{bmatrix} \mathbf{Y}_{1dq} = \mathbf{H}_{dq}\mathbf{Y}_{1dq}. \tag{3.44}$$

This important equation shows that, although we have neglected phase coupling in the a-b-c frame of reference, the d-q components are coupled through the speed voltage term. The matrix **H** defines the nature of this coupling.

Turning now to the RHS of (3.36), we use (3.37) to write

$$\mathbf{f}_{abc} = \mathbf{A}\mathbf{Y}_{1abc} + \mathbf{B}\mathbf{F}_{abc} + \mathbf{B}_1\dot{\mathbf{F}}_{abc}. \tag{3.45}$$

Then

$$\mathbf{f}_{0dq} = \mathbf{P}\mathbf{f}_{abc} = \mathbf{P}\mathbf{A}\mathbf{Y}_{1abc} + \mathbf{P}\mathbf{B}\mathbf{F}_{abc} + \mathbf{P}\mathbf{B}_1\dot{\mathbf{F}}_{abc} \tag{3.46}$$

or

$$\mathbf{f}_{0dq} = \mathbf{P}\mathbf{A}\mathbf{P}^{-1}\mathbf{Y}_{10dq} + \mathbf{P}\mathbf{B}\mathbf{P}^{-1}\mathbf{F}_{0dq} + \mathbf{P}\mathbf{B}_1\mathbf{P}^{-1}\mathbf{P}\dot{\mathbf{F}}_{abc}. \tag{3.47}$$

Now, we note that, because the a-b-c network was assumed to be uncoupled, we may write

$$\begin{aligned} \mathbf{P}\mathbf{A}\mathbf{P}^{-1} &= \mathbf{A} \\ \mathbf{P}\mathbf{B}\mathbf{P}^{-1} &= \mathbf{B} \\ \mathbf{P}\mathbf{B}_1\mathbf{P}^{-1} &= \mathbf{B}_1 \end{aligned} \tag{3.48}$$

since these three system matrices are diagonal.

We further simplify the notation in (3.46) by defining the following new variable. Define

$$\begin{aligned} \mathbf{g}_{0dq} &= \mathbf{P}\dot{\mathbf{F}}_{abc} = \dot{\mathbf{F}}_{0dq} - \dot{\mathbf{P}}\mathbf{P}^{-1}\mathbf{F}_{0dq} \\ &= \dot{\mathbf{F}}_{0dq} - \begin{bmatrix} 0 & 0 & 0 \\ 0 & 0 & -\omega \\ 0 & +\omega & 0 \end{bmatrix} \mathbf{F}_{0dq} = \mathbf{H}\mathbf{F}_{0dq}. \end{aligned} \tag{3.49}$$

Then, ignoring the zero sequence, we write

$$\mathbf{g}_{dq} = \begin{bmatrix} p & +\omega \\ -\omega & p \end{bmatrix} \mathbf{F}_{dq} = \mathbf{H}_{dq}\mathbf{F}_{dq} \tag{3.50}$$

where we observe exactly the same cross coupling that we had in (3.44). Finally, then, (3.46) can be simplified to

$$\mathbf{f}_{0dq} = \mathbf{A}\mathbf{Y}_{10dq} + \mathbf{B}\mathbf{F}_{0dq} + \mathbf{B}_1\mathbf{g}_{0dq}. \tag{3.51}$$

The equation is exactly the same for the *d-q* case with the zero deleted from the notation.

We may use the same technique to show that the second part of (3.36) may be written as

$$\mathbf{Y}_{20dq} = \mathbf{C}\mathbf{Y}_{10dq} + \mathbf{D}\mathbf{F}_{0dq} + \mathbf{D}_1\mathbf{g}_{0dq}. \tag{3.52}$$

In summary, we write the following equations. First we write the uncoupled equations for $\mathbf{f}_{dq}$ and $\mathbf{Y}_{2dq}$,

$$\begin{aligned} \mathbf{f}_{dq} &= \mathbf{A}_{dq}\mathbf{Y}_{1dq} + \mathbf{B}_{dq}\mathbf{F}_{dq} + \mathbf{B}_{1dq}\mathbf{g}_{dq} \\ \mathbf{Y}_{2dq} &= \mathbf{C}_{dq}\mathbf{Y}_{1dq} + \mathbf{D}_{dq}\mathbf{F}_{dq} + \mathbf{D}_{1dq}\mathbf{g}_{dq}. \end{aligned} \tag{3.53}$$

These equations are truly uncoupled, with the *d* variables being functions only of other *d* variables, and similarly for *q* variables.

Second, we write the coupled equations.

$$\begin{aligned} \mathbf{f}_{dq} &= \mathbf{H}_{dq}\mathbf{Y}_{1dq} \\ \mathbf{g}_{dq} &= \mathbf{H}_{dq}\mathbf{F}_{dq}. \end{aligned} \tag{3.54}$$

In these equations, the dependent *d* variables are functions of both *d* and *q* independent variables.

From the above, we may also write

$$\begin{aligned} \dot{y}_{1d} &= f_d + \omega y_{1q} \\ \dot{y}_{1q} &= f_q - \omega y_{1d} \end{aligned} \tag{3.55}$$

which shows the coupling more clearly as coming through the speed voltage terms.

In solving the a-b-c transmission network, a single phase representation is entered and the matrices **A**, **B**, $\mathbf{B_1}$, **C**, **D**, and $\mathbf{D_1}$ formed. These matrices are used twice, once for the d and once for the q axis, to form (3.51) and (3.52). This reduces by a factor of four the size of problem to be solved and permits the solution of much larger systems for a given computer memory size.

Example 3.2 Extend Example 3.1 to write the d-q equations for the system shown in Figure 3.7.

From the previous example, we have the following matrices defined.

$$\begin{aligned} &\mathbf{A} = \mathbf{0} && \mathbf{C} = \mathbf{U} \\ &\mathbf{B} = \mathbf{C}_o^{-1} = \mathbf{S} && \mathbf{D} = \mathbf{R} \\ &\mathbf{B}_1 = \mathbf{0} && \mathbf{D}_1 = \mathbf{L} \end{aligned}$$

where we call the inverse of the capacitance matrix **S**, the elastance matrix. Eliminating the character $\mathbf{C_o}$ from the right-hand side of the above also eliminates the confusion between the two different **C** matrices, one for the state space form and one for capacitance.

Then we may write

$$\mathbf{f}_{0dq} = \mathbf{AY}_{10dq} + \mathbf{BF}_{0dq} + \mathbf{B}_1\mathbf{g}_{0dq} = \mathbf{SF}_{0dq} \tag{3.56}$$

$$\begin{aligned} \mathbf{Y}_{20dq} &= \mathbf{CY}_{10dq} + \mathbf{DF}_{0dq} + \mathbf{D}_1\mathbf{g}_{0dq} \\ &= \mathbf{UY}_{10dq} + \mathbf{RF}_{0dq} + \mathbf{Lg}_{0dq} \end{aligned} \tag{3.57}$$

$$\begin{aligned} \mathbf{g}_{0dq} &= \mathbf{HF}_{0dq} \\ \mathbf{f}_{0dq} &= \mathbf{HY}_{10dq}. \end{aligned} \tag{3.58}$$

Writing out the d- and q-axis equations separately, we have the following:

$$\begin{aligned} \dot{y}_{1d} &= f_d + \omega y_{1q} = S\dot{F}_d + \omega y_{1q} \\ \dot{y}_{1q} &= f_q - \omega y_{1d} = S\dot{F}_q - \omega y_{1d} \end{aligned} \tag{3.59}$$

$$y_{2d} = y_{1d} + RF_d + L\dot{F}_d + \omega LF_q$$
$$y_{2q} = y_{1q} + RF_q + L\dot{F}_q - \omega LF_d \tag{3.60}$$

We can now include these new equations in the state-space model developed in Chapter 2 for the generator and exciter. This will show how the network equations affect the state-space formulation.

From (3.55) and (3.33) we identify the states and write the voltages as generator terminal voltages

$$\dot{v}_d = Spi_d + \omega v_q$$
$$\dot{v}_q = Spi_q - \omega v_d \tag{3.61}$$

where we use the p operator to avoid confusion of having two dots over the letter i. Now, from the generator flux linkage equation (2.129) we compute the derivatives of the currents

$$pi_d = \Gamma_{dd}\dot{\psi}_d + \Gamma_{dF}\dot{\psi}_F + \Gamma_{dD}\dot{\psi}_D$$
$$pi_q = \Gamma_{qq}\dot{\psi}_q + \Gamma_{qG}\dot{\psi}_G + \Gamma_{qQ}\dot{\psi}_Q. \tag{3.62}$$

The flux linkage derivatives are known from the state equations given by (2.144), including the exciter. These derivatives are functions of the state variables and can be substituted into (3.62) with the result substituted into (3.61). This gives the new state equations including the network. These new equations may be written as (they are the 11th and 12th state variables)

$$\dot{v}_d = A_{11-1}\psi_d + A_{11-2}\psi_F + A_{11-3}\psi_D + A_{11-4}\psi_q + A_{11-7}\omega + A_{11-10}E_{FD} + \omega_o v_q - S\Gamma_{dd}\omega_B v_d$$

$$\dot{v}_q = A_{12-1}\psi_d + A_{12-4}\psi_q + A_{12-5}\psi_G + A_{12-6}\psi_Q + A_{12-7}\omega - \omega_o v_d - S\Gamma_{qq}\omega_B v_q \tag{3.63}$$

where

$$A_{8-11} = \frac{v_{do}K_R}{V_{to}\tau_R} \qquad A_{11-1} = S(\Gamma_{dd}h_{dd} + \Gamma_{dF}h_{Fd} + \Gamma_{dD}h_{Dd})$$

$$A_{8-12} = \frac{v_{qo}K_R}{V_{to}\tau_R} \qquad A_{11-2} = S(\Gamma_{dd}h_{dF} + \Gamma_{dF}h_{FF} + \Gamma_{dD}h_{DF})$$

$$A_{11-3} = S(\Gamma_{dd}h_{dD} + \Gamma_{dF}h_{FD} + \Gamma_{dD}h_{DD})$$

$$A_{11-10} = S\Gamma_{dF}k_x \qquad A_{11-4} = -S\Gamma_{dd}\omega_B\omega_o$$
$$A_{11-11} = -S\Gamma_{dd}\omega_B \qquad A_{11-7} = -S\Gamma_{dd}\omega_B\psi_{qo} + v_{qo}$$
$$A_{12-1} = +S\Gamma_{qq}\omega_B\omega_o \qquad A_{12-4} = S\left(\Gamma_{qq}h_{qq} + \Gamma_{qG}h_{Gq} + \Gamma_{qQ}h_{Qq}\right)$$
$$A_{12-7} = +S\Gamma_{qq}\omega_B\psi_{do} - v_{do} \qquad A_{12-5} = S\left(\Gamma_{qq}h_{qG} + \Gamma_{qG}h_{GG} + \Gamma_{qQ}h_{QG}\right)$$
$$A_{12-12} = -S\Gamma_{qq}\omega_B \qquad A_{12-6} = S\left(\Gamma_{qq}h_{qQ} + \Gamma_{qG}h_{GQ} + \Gamma_{qQ}h_{QQ}\right) \tag{3.64}$$

Substituting these coefficients into the state space equations we write (3.65), which is shown on the following page, where we also define the following matrix elements:

$$A_{7-1} = \frac{I_{dd}}{2H} \qquad A_{7-5} = \frac{-I_{qG}}{2H}$$
$$A_{7-2} = \frac{I_{dF}}{2H} \qquad A_{7-6} = \frac{-I_{qQ}}{2H}$$
$$A_{7-3} = \frac{I_{dD}}{2H} \qquad A_{7-7} = \frac{-D}{2H}$$
$$A_{7-4} = \frac{-I_{qq}}{2H} \qquad B_{7-1} = \frac{1}{2H}$$

This completes Example 3.2.

3.6 GENERATOR FREQUENCY TRANSFORMATION

If there is more than one generator connected to the system, the question arises as to which angular velocity should be used in the various equations for the models of the machines and the transmission network. In this section we propose that the network be considered to be operating with a synchronous reference frame, and that each machine is operating with a reference frame connected to its rotor. At the point where the machine connects to the system, a transformation will be introduced to account for the motion of the machine reference frame with respect to the system. Figure 3.8 shows the relationship between the system reference frame, noted as *D-Q* and the rotor reference frame of the given machine, noted as *d-q*.

For simplicity, let the subscript "*s*" represent the system reference frame and the subscript "*m*" represent the machine, with δ being the angle

$$
\begin{bmatrix} \dot{\psi}_d \\ \dot{\psi}_F \\ \dot{\psi}_D \\ \dot{\psi}_q \\ \dot{\psi}_G \\ \dot{\psi}_Q \\ \dot{\omega} \\ \dot{V}_1 \\ \dot{V}_2 \\ \dot{E}_{FD} \\ \dot{v}_d \\ \dot{v}_q \end{bmatrix}
=
\left[\begin{array}{ccc|ccc|c|ccc|cc}
h_{dd} & h_{dF} & h_{dD} & -\omega_B\omega_o & 0 & 0 & -\omega_B\psi_o & 0 & 0 & 0 & -\omega_B & 0 \\
h_{Fd} & h_{FF} & h_{FD} & 0 & 0 & 0 & 0 & 0 & 0 & k_x & 0 & 0 \\
h_{Dd} & h_{DF} & h_{DD} & 0 & 0 & 0 & 0 & 0 & 0 & 0 & 0 & 0 \\
\hline
\omega_B\omega_o & 0 & 0 & h_{qq} & h_{qG} & h_{qQ} & \omega_B\psi_o & 0 & 0 & 0 & 0 & -\omega_B \\
0 & 0 & 0 & h_{Gq} & h_{GG} & h_{GQ} & 0 & 0 & 0 & 0 & 0 & 0 \\
0 & 0 & 0 & h_{Qq} & h_{QG} & h_{QQ} & 0 & 0 & 0 & 0 & 0 & 0 \\
\hline
A_{7-1} & A_{7-2} & A_{7-3} & A_{7-4} & A_{7-5} & A_{7-6} & A_{7-7} & 0 & 0 & 0 & 0 & 0 \\
\hline
0 & 0 & 0 & 0 & 0 & 0 & 0 & A_{8-8} & 0 & 0 & A_{8-11} & A_{8-12} \\
0 & 0 & 0 & 0 & 0 & 0 & 0 & A_{9-8} & A_{9-9} & 0 & 0 & 0 \\
0 & 0 & 0 & 0 & 0 & 0 & 0 & A_{10-8} & A_{10-9} & A_{10-10} & 0 & 0 \\
\hline
A_{11-1} & A_{11-2} & A_{11-3} & A_{11-4} & 0 & 0 & A_{11-7} & 0 & 0 & A_{11-10} & A_{11-11} & \omega_o \\
A_{12-1} & 0 & 0 & A_{12-4} & A_{12-5} & A_{12-6} & A_{12-7} & 0 & 0 & 0 & -\omega_o & A_{12-12}
\end{array}\right]
\begin{bmatrix} \psi_d \\ \psi_F \\ \psi_D \\ \psi_q \\ \psi_G \\ \psi_Q \\ \omega \\ V_1 \\ V_2 \\ E_{FD} \\ v_d \\ v_q \end{bmatrix}
+
\left[\begin{array}{ccc}
0 & 0 & 0 \\
0 & 0 & 0 \\
0 & 0 & 0 \\
\hline
0 & 0 & 0 \\
0 & 0 & 0 \\
0 & 0 & 0 \\
\hline
B_{7-1} & 0 & 0 \\
\hline
0 & 0 & 0 \\
0 & B_{9-2} & B_{9-3} \\
0 & B_{10-2} & B_{10-3} \\
\hline
0 & 0 & 0 \\
0 & 0 & 0
\end{array}\right]
\begin{bmatrix} T_m \\ V_{REF} \\ V_S \end{bmatrix}
\tag{3.65}
$$

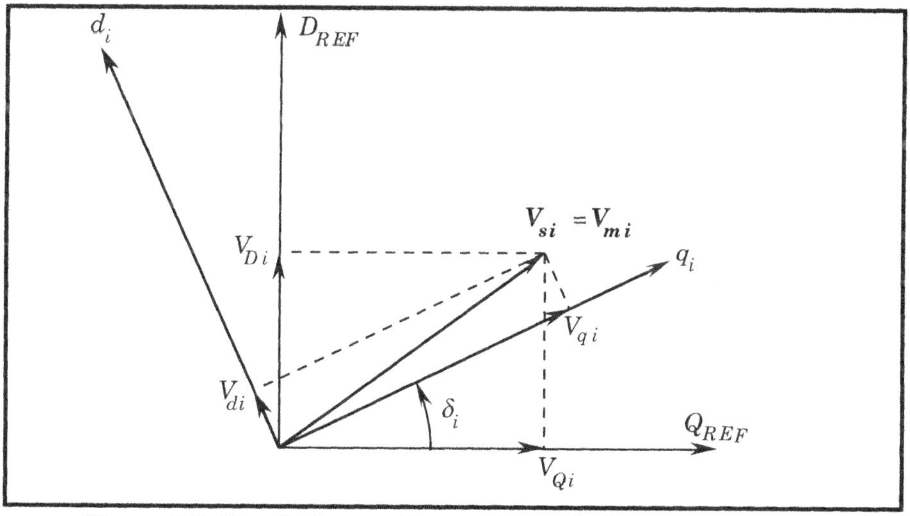

Figure 3.8 Two Reference Frames for a Machine Quantity [9]

between the two, as shown in Figure 3.8. It is easily shown that the relationship between the two may be written as [9]

$$\boldsymbol{V}_{si} = \boldsymbol{V}_{mi} e^{j\delta_i}. \tag{3.66}$$

Stated in terms of the d and q components of the voltages, we may write

$$\begin{bmatrix} V_{dm} \\ V_{qm} \end{bmatrix} = \begin{bmatrix} \cos\delta & -\sin\delta \\ \sin\delta & \cos\delta \end{bmatrix} \begin{bmatrix} V_{ds} \\ V_{qs} \end{bmatrix} \tag{3.67}$$

$$\begin{bmatrix} V_{ds} \\ V_{qs} \end{bmatrix} = \begin{bmatrix} \cos\delta & \sin\delta \\ -\sin\delta & \cos\delta \end{bmatrix} \begin{bmatrix} V_{dm} \\ V_{qm} \end{bmatrix}. \tag{3.68}$$

These equations must be linearized in order to use them in the SSR analysis programs, such as eigenvalue analysis. The machine voltage in terms of the system voltage would be given as follows.

$$\begin{bmatrix} \Delta V_{dm} \\ \Delta V_{qm} \end{bmatrix} = \begin{bmatrix} \cos\delta_o & -\sin\delta_o \\ \sin\delta_o & \cos\delta_o \end{bmatrix} \begin{bmatrix} \Delta V_{ds} \\ \Delta V_{qs} \end{bmatrix} - \begin{bmatrix} \sin\delta_o & \cos\delta_o \\ -\cos\delta_o & \sin\delta_o \end{bmatrix} \begin{bmatrix} V_{dso} \\ V_{qso} \end{bmatrix} \Delta\delta. \tag{3.69}$$

The system injection current in terms of the machine injection current is given by

$$\begin{bmatrix} \Delta I_{ds} \\ \Delta I_{qs} \end{bmatrix} = \begin{bmatrix} \cos\delta_o & \sin\delta_o \\ -\sin\delta_o & \cos\delta_o \end{bmatrix} \begin{bmatrix} \Delta I_{dm} \\ \Delta I_{qm} \end{bmatrix} - \begin{bmatrix} \sin\delta_o & -\cos\delta_o \\ \cos\delta_o & \sin\delta_o \end{bmatrix} \begin{bmatrix} I_{dmo} \\ I_{qmo} \end{bmatrix} \Delta\delta. \tag{3.70}$$

Since the derivative of the injection current is also needed, we compute this quantity by computing the derivative of (3.68), with the equation written for current instead of voltage.

$$\begin{bmatrix} \dot{I}_{ds} \\ \dot{I}_{qs} \end{bmatrix} = \begin{bmatrix} \cos\delta & \sin\delta \\ -\sin\delta & \cos\delta \end{bmatrix} \begin{bmatrix} \dot{I}_{dm} \\ \dot{I}_{qm} \end{bmatrix} - \begin{bmatrix} \sin\delta & -\cos\delta \\ \cos\delta & \sin\delta \end{bmatrix} \begin{bmatrix} I_{dm} \\ I_{qm} \end{bmatrix} \omega. \tag{3.71}$$

This expression is linearized to compute

$$\begin{bmatrix} \Delta\dot{I}_{ds} \\ \Delta\dot{I}_{qs} \end{bmatrix} = \begin{bmatrix} \cos\delta_o & \sin\delta_o \\ -\sin\delta_o & \cos\delta_o \end{bmatrix} \begin{bmatrix} \Delta\dot{I}_{dm} \\ \Delta\dot{I}_{qm} \end{bmatrix} - \begin{bmatrix} \sin\delta_o & -\cos\delta_o \\ \cos\delta_o & \sin\delta_o \end{bmatrix} \begin{bmatrix} I_{dmo} \\ I_{qmo} \end{bmatrix} \Delta\omega \tag{3.72}$$

where we assume that the derivative of δ and the derivatives of the currents are zero in the steady state.

3.7 MODULATION OF THE 60 Hz NETWORK RESPONSE

The response of any underdamped electric network containing R-L-C elements will include two components in the time domain current waveforms:

1. A sinusoidal component at the frequency of the driving voltage.
2. A damped sinusoidal component at a frequency that depends entirely on the elements of the network.

This can be shown by considering a simple R-L-C series connected network branch, across which we apply a fundamental frequency voltage

$$v(t) = \sqrt{2}V\sin(\omega_1 t + \theta). \tag{3.73}$$

This voltage has the Laplace transform

$$V(s) = \sqrt{2}V \frac{s\sin\theta + \omega_1 \cos\theta}{s^2 + \omega_1^2}. \tag{3.74}$$

Now the impedance of the series branch in the Laplace domain is given by

$$Z(s) = R + sL + \frac{1}{sC} \tag{3.75}$$

and we solve for the current in the branch as

$$I(s) = \frac{V(s)}{Z(s)} = \frac{V(s)}{R + sL + \frac{1}{sC}} = \frac{\frac{sV(s)}{L}}{s^2 + \frac{R}{L}s + \frac{1}{LC}}. \tag{3.76}$$

It is common practice to write this result in one of the following forms:

$$I(s) = \frac{\frac{sV(s)}{L}}{s^2 + 2\zeta\omega_n s + \omega_n^2} = \frac{\frac{sV(s)}{L}}{(s-a)^2 + \omega_2^2} \tag{3.77}$$

where we define the following:

1. The Undamped Natural Frequency

$$\omega_n = \sqrt{\frac{1}{LC}} \tag{3.78}$$

2. The Damping Ratio

$$\zeta = \frac{R}{2}\sqrt{\frac{C}{L}} \tag{3.79}$$

3. The Damping Rate

$$a = -\zeta\omega_n \tag{3.80}$$

4. The Damped Frequency

$$\omega_2 = \omega_n \sqrt{1-\zeta^2}. \tag{3.81}$$

Substituting the voltage transform of (3.74) into (3.77) and solving for the current, we compute

$$i(t) = K\left[A\sin(\omega_1 t + \psi_1) + Be^{-\zeta\omega_2 t}\sin(\omega_2 t + \psi_2)\right] \tag{3.82}$$

where

$$K = \frac{\sqrt{2}\omega_2 CV\cos\theta}{\left[\left(\omega_2^2 - \omega_1^2\right) + 4\zeta^2\omega_1^2\omega_2^2\right]^2} \tag{3.83}$$

$$A = \omega_2\sqrt{1+a^2\omega_1^2} \tag{3.84}$$

$$B = \omega_1\sqrt{\frac{1-2a\zeta\omega_2 + a^2\omega_2^2}{1-\zeta^2}} \tag{3.85}$$

$$\psi_1 = \theta + \tan^{-1}\left(\frac{2\zeta\omega_1\omega_2}{\omega_2^2 - \omega_1^2}\right) \tag{3.86}$$

$$d = \sqrt{1-\zeta^2} \tag{3.87}$$

and

$$\psi_2 = \tan^{-1}\left(\frac{ad\omega_2}{1-a\zeta\omega_2}\right) - \tan^{-1}\left(\frac{-2d\zeta\omega_2^2}{2\zeta\omega_2^2 + \omega_1^2 - \omega_2^2}\right). \tag{3.88}$$

Thus, we see two frequencies in the current response

ω_1 = The Frequency of the Driving Voltage

ω_2 = A Frequency Depending on the Network Elements

where

$$\omega_2 = d\omega_n = \frac{1}{2L}\left(\frac{4L - R^2C}{C}\right). \tag{3.89}$$

Now, suppose that the foregoing result is the solution for only phase a of a balanced three-phase network. If we solve for the phase b and c currents, they will also have the two frequencies present, but with different coefficients on the transient response component. If these three currents are processed through a Park's transformation, the current waveforms are both multiplied by the transcendental functions $\cos\theta$, $\sin\theta$, and similar terms displaced by ±120 degrees.

We know that the Park's transformation of the balanced system frequency currents will give constant (dc) currents in their d and q axis components. But how will the transient components be transformed? These transformations lead to many terms like

$$\cos\,\theta\,\sin\,\omega_2 t \tag{3.90}$$

where it is recalled that

$$\theta = \omega_1 t + \zeta + \frac{\pi}{2} \tag{3.91}$$

Now, it is easily shown that

$$\begin{aligned}\cos\theta\sin\omega_2 t &= \frac{1}{2}\left[\sin(\theta+\omega_2 t) - \sin(\theta-\omega_2 t)\right] \\ &= \frac{1}{2}\left\{\cos\left[(\omega_1+\omega_2)t+\zeta\right] - \cos\left[(\omega_1-\omega_2)t+\right]\right\}\zeta\end{aligned} \tag{3.92}$$

Thus the transient network currents appear to the generator rotor as currents with frequencies of the sum and difference of the two frequencies that occur in the network currents. This leads to supersynchronous and subsynchronous torques, the latter of which are of particular interest in SSR studies. These subsynchronous currents inject energy into the rotating mass of the shaft that may be very close to a natural mechanical resonance. Even very small energies injected at these resonant frequencies can lead to substantial oscillatory motion and resultant large shaft stresses. Since these oscillatory currents in the network are poorly damped, the subsynchronous currents may persist long enough to cause permanent shaft fatigue and possible failure.

In computing the eigenvalues of a system with series compensated transmission lines, we find natural frequencies associated with each R-L-C

line section as noted above. These frequencies appear in the computed eigenvalues as sum and difference quantities with respect to the system base frequency.

3.8 REFERENCES FOR CHAPTER 3

1. Rohrer, Ronald A., **Circuit Theory: An Introduction to the State Variable Approach**, McGraw-Hill, New York, 1970.

2. Karni, Shlomo, **Intermediate Network Analysis**, Allyn and Bacon, Boston, 1971.

3. Pottle, C., "State-Space Techniques for General Active Network Analysis," Chapter 3 of **System Analysis by Digital Computers**, F. F. Kuo and J. F. Kaiser, Eds., John Wiley and Sons, New York, 1966, pp. 59-98.

4. Chen, Wai-Kai, **Linear Networks and Systems**, Brooks/Cole Engineering Division, Wadsworth, Belmont, California, 1983.

5. Wing, Omar, **Circuit Theory with Computer Methods**, Hemisphere Pub. Corp., Washington, 1978.

6. Dervisoglu, A., "State Models of Active RLC Networks," Report R-237, Coordinated Science Laboratories, University of Illinois, December 1964.

7. Balabanian, N. and T. Bickart, **Linear Network Theory**, Matrix Publishers, Inc., Beaverton, Oregon, 1981, pp. 118-132.

8. Pottle, C., "A 'Textbook' Computerized State-Space Network Analysis Algorithm, **IEEE Trans. on Circuit Theory**, v. CT-16, November 1969, pp. 566-568.

9. Anderson, P. M., and A. A. Fouad, **Power System Control and Stability**, Iowa State University Press, Ames, Iowa, 1977.

CHAPTER 4

THE TURBINE-GENERATOR SHAFT MODEL

This chapter presents the equations of the turbine-generator shaft and the derivation of a general shaft model for use in SSR eigenvalue computations. The shaft model assumes that the shaft may be divided into finite inertia segments, i.e., a lumped spring-mass model. Actual shafts consist of connected continuous machined steel cylinders, to which massive elements are attached. Certain studies may require a more detailed model of the shaft than that derived here. The manufacturers should be consulted as to the need for such highly detailed models. The manufacturers must also be consulted for data on any shaft model. Shaft model data are discussed further in Chapter 6.

In eigenvalue calculations, we are concerned with small perturbations from the normal operating mode. For the turbine-generator shaft, this means that the material is stressed within the elastic limit of that material and that linear relationships may be written according to Hooke's law of material deformation and Newton's law of mechanics. The material behavior is characterized by equivalent parameters that relate the average spring constant and damping between lumped masses on the shaft and the damping or viscous friction between masses and the stationary frame.

4.1 DEFINITIONS AND CONVENTIONS

The shaft physical model is shown in Figure 4.1. We arbitrarily number the turbines from one to "n", and connect adjacent turbines by a spring of constant K and a dashpot of damping D that represents the average material behavior between masses. Another dashpot is represented from each turbine mass to the fixed reference frame to represent damping due to the relative motion of the turbine blades (sometimes called steam damping). Each turbine provides an accelerating torque T, shown as positive in the counter-clockwise direction viewed looking at the shaft from the left end. A positive acceleration due to this torque results in an angular acceleration in the counter-clockwise direction and in an instantaneous angular velocity as shown in Figure 4.1.

The generator is denoted by the subscript "g" in Figure 4.1 with its torque T_e shown as positive in a direction opposite to the driving torque of the turbines. This means that the electromagnetic torque of the generator is

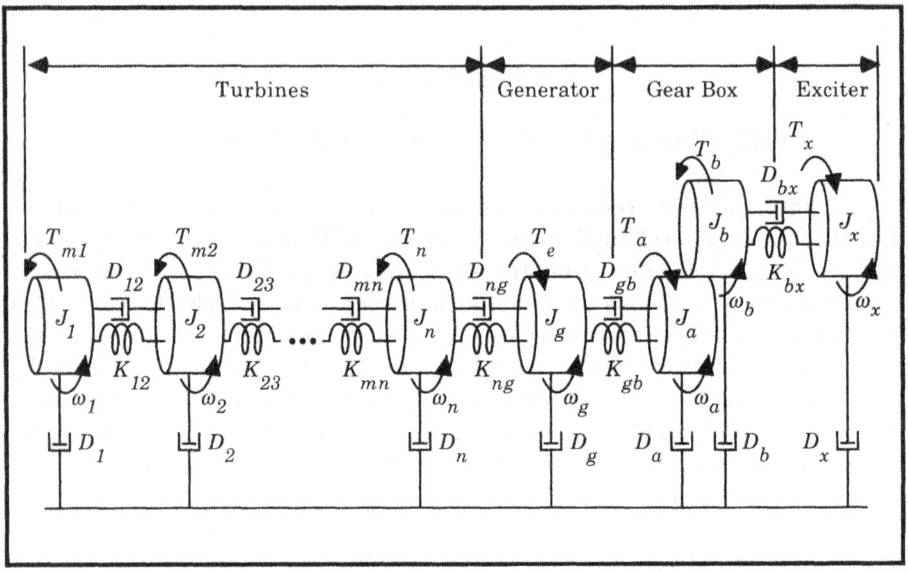

Figure 4.1 Lumped Spring-Mass Model of a Turbine-Generator Shaft

actually a load or retarding torque to the turbines. The generator angular velocity is measured as a positive quantity in the same direction as the turbines, namely, counter-clockwise.

A shaft-mounted exciter is also represented at the far right end of the shaft of Figure 4.1, and is denoted by the subscript "x". The exciter torque T_x is also represented as a load torque.

A provision is made for a gear box between the generator and the exciter, a feature that is found in some turbine-generator systems. The gear box is illustrated as having a moments of inertia J_a and J_b, although these may be negligible inertias on certain geared systems. No load torque is shown for the gear box on the assumption that any losses in the gearing are very small compared to the rated torques under consideration.

The physical model depicted in Figure 4.1 is commonly used in the study of turbine-generator shaft dynamics. The turbine lumped masses are usually representative of the major turbine sections, such as the high-pressure turbine, intermediate-pressure turbine, and the various low-pressure turbines. The manufacturers are able to supply data on these various turbine sections as well as the entire shaft system.

Two conventions are used in this chapter that require explanation. Figure 4.1 shows the various shaft masses represented by lumped moments of inertia J for each mass. This notation is the ANSI standard notation for moment of inertia and is a common notation. This notation, however, presents a problem because it is difficult to distinguish between the moment of inertia and the rotating energy in joules, also commonly depicted by the unit symbol abbreviation "J", which is also an ANSI standard. Technically, the unit symbol abbreviation is specified by the standards to be shown in Roman typeface and the moment of inertia, or other variables, in italic typeface. We resolve this problem by always using "J" to represent the moment of inertia and shall use the entire word "Joule" when referring to the energy.

Another notational complication that will be encountered in this section is the reference to both mechanical and electrical angular velocities, which are related by the number of poles "p" in the generator design. In previous sections, however, we have sometimes used "p" to represent the derivative with respect to time. This problem is resolved by writing the differential equations for the shaft in the Laplace domain, using the Laplace variable "s" rather than the time derivative "p".

Hence, for this chapter the letter "p" is reserved to mean the number of poles in the generator. Much of the world uses this same symbol to represent the number of pairs of poles, a practice that is not common in the North America. Our approach will be to use p for the *number* of poles.

The torque equations for the shaft lumped spring-mass sections are the fundamental relations that determine the dynamic performance. These equations are written in agreement with Newton's second law for rotating bodies. For example, for the jth mass, we may write the following equation, assuming that the jth mass is connected by elastic shaft sections to masses i and k.

$$
\begin{aligned}
J_j \frac{d\omega_j}{dt} &= T_j - D_{ji}(\omega_j - \omega_i) - K_{ji}(\theta_j - \theta_i) \\
&\quad - D_{jk}(\omega_j - \omega_k) - K_{jk}(\theta_j - \theta_k)
\end{aligned} \tag{4.1}
$$

where the terms in (4.1) are further defined in Table 4.1.

These terms will be used in the development of equations for the turbine-generator shaft model shown in Figure 4.1

Table 4.1 Description of Terms for the Torque Equation

Item	**Symbol**	**Units**	**Dimensions** [VIT System]
Moment of Inertia	J	$kg.m^2$	$[VIT^3]$
Torque	T	N.m or Joules	[VIT]
Damping Coefficient	D	Joule-s/rad	$[VIT^2]$
Spring Constant	K	Joule/rad or M.m/rad	[VIT]
Angular Velocity	ω	rad/s	$[T^{-1}]$
Angle	θ	rad	– – –
Time	t	s	[T]
Power	P	W	[VI]
Energy	W	Joule	[VIT]
Laplace	s	s^{-1}	$[T^{-1}]$

4.2 THE SHAFT TORQUE EQUATIONS

This section presents the shaft torque equations for the shaft configuration shown in Figure 4.1. All equations are written in mks units using Newton's Laws of mechanics for a rotational system, which gives shaft torques in newton-meters (N.m). We also define the gear ratio of the gear box, shown in Figure 4.2, as follows.

$$\text{Gear Ratio} = R_G = \frac{\omega_a}{\omega_b} > 1 \tag{4.2}$$

where

ω_a = Angular Velocity of Gear Box on Generator Side
ω_b = Angular Velocity of Gear Box on Exciter Side

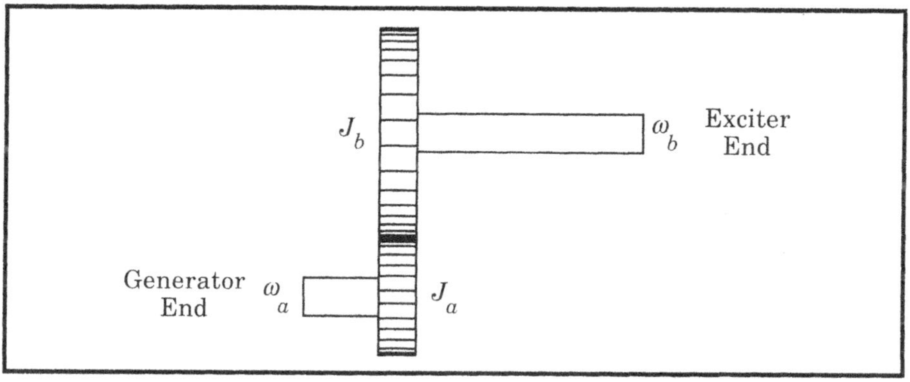

Figure 4.2 Shaft Gear Box Between Generator and Exciter

where both angular velocities are in *mechanical* radians per second rather than *electrical* radians per second, which is often used for the generator speed.

The shaft torques are determined by applying Newton's second law for a rotational system to compute the torques in N.m. The general form of this equation is given, in the Laplace domain, by

$$J\frac{d\omega}{dt} = \sum Torques. \tag{4.3}$$

The right-hand side of (4.3) is a collection of terms that represent externally applied torques as well as retarding torques due to reactions within the shaft material, such as resilience and damping, with torques that tend to accelerate the shaft in the counter-clockwise direction taken as positive.

Working through the shaft inertias from left to right in Figure 4.1, we write the following equations.

$$\begin{aligned}
J_1\dot{\omega}_1 &= T_{m1} - D_1\omega_1 - D_{12}(\omega_1 - \omega_2) - K_{12}(\theta_1 - \theta_2) \\
J_2\dot{\omega}_2 &= T_{m2} - D_2\omega_2 - D_{21}(\omega_2 - \omega_1) - K_{21}(\theta_2 - \theta_1) \\
&\quad - D_{23}(\omega_2 - \omega_3) - K_{23}(\theta_2 - \theta_3) \\
&\dots \\
J_n\dot{\omega}_n &= T_{mn} - D_n\omega_n - D_{nm}(\omega_n - \omega_m) - K_{nm}(\theta_n - \theta_m) \\
&\quad - D_{ng}(\omega_n - \omega_g) - K_{ng}(\theta_n - \theta_g)
\end{aligned}$$

$$
\begin{aligned}
J_g\dot{\omega}_g &= -T_e - D_g\omega_g - D_{gn}(\omega_g - \omega_n) - K_{gn}(\theta_g - \theta_n) \\
&\qquad - D_{ga}(\omega_g - \omega_a) - K_{ga}(\theta_g - \theta_a) \\
J_a\dot{\omega}_a &= T_a - D_a\omega_a - D_{ag}(\omega_a - \omega_g) - K_{ag}(\theta_a - \theta_g) \\
J_b\dot{\omega}_b &= -T_b - D_b\omega_b - D_{bx}(\omega_b - \omega_x) - K_{bx}(\theta_b - \theta_x) \\
J_x\dot{\omega}_x &= -T_x - D_x\omega_x - D_{xb}(\omega_x - \omega_b) - K_{xb}(\theta_x - \theta_b).
\end{aligned}
\tag{4.4}
$$

For the gear train we may write

$$
R_G = \frac{\omega_a}{\omega_b} = \frac{N_b}{N_a} = \frac{T_b}{T_a}
\tag{4.5}
$$

where

T_a = Load Torque Transmitted to Shaft a
T_b = Load Torque Transmitted to Shaft b
N_a = Number of Teeth in Gear a
N_b = Number of Teeth in Gear b.

In Figure 4.1 and (4.4) we let $m = n - 1$ identify the mass just left of mass n. Equations 4.4 and 4.5 completely determine the dynamic behavior of the shaft, and the eigenvalues associated with these equations provide data on the various response frequencies and modes of oscillation of the shaft system.

It is conceivable that one could model the material resilience of the meshed teeth of the gear train by an appropriate spring constant. Tests conducted by equipment manufacturers indicate that such a spring constant would appear to be very large at the low frequencies under consideration in most torsional interaction studies [2]. There is no plausible reason to suspect the presence of any appreciable damping in the gear itself. If such damping exists, it is small and is usually neglected. Backlash in the gear is an important phenomenon when the shaft is unloaded, but is not important in studies of the loaded machine, such as those conducted for SSR.

Geared turbine-generator shafts are not common in newer machines. They are mostly confined to older units with dc generator exciters. When there is no gear box, the model is unchanged except that the gear ratio is then set to unity.

We now eliminate one of the gear box equations and replace the two equations, designated a and b, by one equation representing an equivalent spring-mass model of the gear system. From (4.5) we write

$$T_b = R_G T_a \tag{4.6}$$

The load torques for each shaft can be solved from the a and b equations of (4.4) and substituted into (4.6) with the result

$$\begin{aligned}\left(R_G^2 J_a + J_b\right) s\omega_B = -\left(R_G^2 D_a + D_b\right)\omega_B - R_G D_{ag}\left(\omega_a - \omega_g\right) \\ -R_G K_{ag}\left(\theta_a - \theta_g\right) - D_{bx}\left(\omega_b - \omega_x\right) - K_{bx}\left(\theta_b - \theta_x\right).\end{aligned} \tag{4.7}$$

Now, define the equivalent gear box quantities

$$\begin{aligned}J_q &= R_G^2 J_a + J_b \\ D_q &= R_G^2 D_a + D_b\end{aligned} \tag{4.8}$$

to write

$$\begin{aligned}J_q s\omega_b = 0 - D_q\omega_b - R_G D_{ag}\left(\omega_a - \omega_g\right) - R_G K_{ag}\left(\theta_a - \theta_g\right) \\ - D_{bx}\left(\omega_b - \omega_x\right) - K_{bx}\left(\theta_b - \theta_x\right).\end{aligned} \tag{4.9}$$

It is also convenient to rewrite (4.4) entirely in terms of the shaft angular velocities. To do this we note that the shaft angular position and velocity are related by the equation

$$\omega = s\theta \tag{4.10}$$

or the speed is the derivative of the angular position. Thus shaft angle is the integral of shaft speed. Making this substitution in (4.4) and incorporating (4.9) we write the following equations:

$$\begin{aligned}J_1 s\omega_1 &= T_{m1} - D_1\omega_1 - \left(D_{12} + \frac{K_{12}}{s}\right)\left(\omega_1 - \omega_2\right) \\ J_2 s\omega_2 &= T_{m2} - D_2\omega_2 - \left(D_{21} + \frac{K_{21}}{s}\right)\left(\omega_2 - \omega_1\right) \\ &\quad - \left(D_{23} + \frac{K_{23}}{s}\right)\left(\omega_2 - \omega_3\right) \\ &\cdots \\ J_n s\omega_2 n &= T_{mn} - D_n\omega_n + \left(D_{mn} + \frac{K_{mn}}{s}\right)\left(\omega_m - \omega_n\right) \\ &\quad - \left(D_{ng} + \frac{K_{ng}}{s}\right)\left(\omega_n - \omega_g\right)\end{aligned}$$

$$
\begin{aligned}
J_g s\omega_g &= -T_e - D_g\omega_g + \left(D_{ng} + \frac{K_{ng}}{s}\right)(\omega_n - \omega_g) \\
&\quad - \left(D_{ga} + \frac{K_{ga}}{s}\right)(\omega_g - \omega_a) \\
J_q s\omega_b &= 0 - D_q\omega_b + R_G\left(D_{ga} + \frac{K_{ga}}{s}\right)(\omega_g - \omega_a) \\
&\quad - \left(D_{bx} + \frac{K_{bx}}{s}\right)(\omega_b - \omega_x) \\
J_x s\omega_x &= -T_x - D_x\omega_x + \left(D_{bx} + \frac{K_{bx}}{s}\right)(\omega_b - \omega_x)
\end{aligned}
\tag{4.11}
$$

It is important to note that the shaft has two rated angular velocities, one on each side of the gear box. We define these rated angular velocities as follows:

$$
\begin{aligned}
\omega_R &= \text{Rated Turbine-Generator Angular Velocity in mechanical rad/s} \\
\omega_{XR} &= \text{Rated Exciter Angular Velocity in mechanical rad/s}
\end{aligned}
\tag{4.12}
$$

Then

$$\omega_R = R_G\omega_{XR}. \tag{4.13}$$

4.3 THE SHAFT POWER EQUATIONS

Steam and hydraulic turbines provide the motive power for driving the turbine shaft. The performance of these devices are best described in terms of the mechanical power output of the turbine rather than the torque. Therefore, we modify the shaft equations from torque to power by the fundamental relation

$$P = T\omega \tag{4.14}$$

which states that each torque equation should be multiplied by the angular velocity of that rotating mass to convert the torque equation to a power equation. Incrementally, we write (4.14) as

$$\Delta P = T_o\,\Delta\omega + \omega_o\,\Delta T \tag{4.15}$$

where the initial angular velocity is assumed to be the steady state rated velocities at each end of the shaft, and these rated velocities are related as noted in (4.13).

Now the basic Newton's law equation expresses the accelerating torque to the product of the moment of inertia and angular acceleration, which we may write in general terms as

$$T_a = Js\omega. \tag{4.16}$$

This expression is true for each mass, as given by (4.11). This general equation for accelerating torque can be written incrementally as

$$\Delta T_a = Js\Delta\omega. \tag{4.17}$$

Then, from (4.15), we write in incremental accelerating power as

$$\Delta P_a = T_{ao}\ \Delta\omega + \omega_o\ \Delta T_a. \tag{4.18}$$

If the system is considered to be in an initial steady-state condition, then the initial accelerating torque is always zero, or

$$\Delta T_{ao} = 0 \tag{4.19}$$

and we may write the incremental accelerating power as

$$\begin{aligned} \Delta P_a &= \omega_o \Delta T_a \\ &= \omega_R \Delta T_a \qquad \text{Generator End} \\ &= \omega_{XR} \Delta T_a \qquad \text{Exciter End.} \end{aligned} \tag{4.20}$$

The procedure for converting the incremental torque equations to power equations, then, is to multiply each torque equation by the rated angular velocity for each lumped inertia, as given by the (4.11). Since these rated velocities are constants the equations remain linear (and Laplace transformable). This means that we multiply all equations except the last two by the rated turbine-generator speed. The last two equations are multiplied by the rated exciter shaft speed. The only restriction to this procedure is that we assume that each shaft equation is initially in equilibrium, i.e., without initial acceleration. Performing the required operation, we write

$$
\begin{aligned}
J_1\omega_R s\omega_1 &= \omega_R T_{m1} - \omega_R D_1\omega_1 - \omega_R\left[D_{12} + \frac{K_{12}}{s}\right](\omega_1 - \omega_2) \\
J_2\omega_R s\omega_2 &= \omega_R T_{m2} - \omega_R D_2\omega_2 + \omega_R\left[D_{12} + \frac{K_{12}}{s}\right](\omega_1 - \omega_2) \\
&\qquad - \omega_R\left[D_{23} + \frac{K_{23}}{s}\right](\omega_2 - \omega_3) \\
&\cdots \\
J_n\omega_R s\omega_n &= \omega_R T_{mn} - \omega_R D_n\omega_n + \omega_R\left[D_{mn} + \frac{K_{mn}}{s}\right](\omega_m - \omega_n) \\
&\qquad - \omega_R\left[D_{ng} + \frac{K_{ng}}{s}\right](\omega_n - \omega_g) \\
J_g\omega_R s\omega_g &= -\omega_R T_e - \omega_R D_g\omega_g + \omega_R\left[D_{ng} + \frac{K_{ng}}{s}\right](\omega_n - \omega_g) \\
&\qquad - \omega_R\left[D_{ga} + \frac{K_{ga}}{s}\right](\omega_g - \omega_a) \\
&\cdots \\
J_q\omega_{XR} s\omega_b &= 0 - \omega_{XR} D_q\omega_b + \omega_{XR}\left[D_{ga} + \frac{K_{ga}}{s}\right](\omega_g - \omega_a) \\
&\qquad - \omega_{XR}\left[D_{bx} + \frac{K_{bx}}{s}\right](\omega_b - \omega_x) \\
J_x\omega_{XR} s\omega_x &= -\omega_{XR} T_x - \omega_{XR} D_x\omega_x + \omega_{XR}\left[D_{bx} + \frac{K_{bx}}{s}\right](\omega_b - \omega_x).
\end{aligned}
\tag{4.21}
$$

We now define the following powers, all at rated speed:

$$
\begin{aligned}
P_{m1} &= \omega_R T_{m1} = \text{Turbine 1 Output Power} \\
P_{m2} &= \omega_R T_{m2} = \text{Turbine 2 Output Power} \\
&\cdots \\
P_{mn} &= \omega_R T_{mn} = \text{Turbine n Output Power} \\
P_e &= \omega_R T_e = \text{Generator Electromagnetic Power} \\
P_x &= \omega_{XR} T_x = \text{Exciter Electromagnetic Power}
\end{aligned}
\tag{4.22}
$$

We also introduce the kinetic energy associated with a given rotating mass, which is defined by the equation

$$\text{Kinetic Energy} = W_k = \frac{1}{2}J\omega^2 \tag{4.23}$$

Using this equation, we may define the kinetic energy at rated speed for each rotating mass on the shaft.

$$\begin{aligned}
W_{ki} &= \frac{1}{2}J_i\omega_R^2, \qquad i = 1,2,\ldots,n,g \\
W_{kq} &= \frac{1}{2}J_q\omega_{XR}^2 \\
W_{kx} &= \frac{1}{2}J_x\omega_{XR}^2.
\end{aligned} \tag{4.24}$$

Thus, the coefficients of the left-hand side of each equation in (4.21) can be expressed in terms of the kinetic energy of that rotating mass. Rewriting (4.21) in terms of power and kinetic energy, we have the following.

$$\begin{aligned}
\frac{2W_{k1}}{\omega_R}s\omega_1 &= P_{m1} - \omega_R D_1\omega_1 - \omega_R\left[D_{12} + \frac{K_{12}}{s}\right](\omega_1 - \omega_2) \\
\frac{2W_{k2}}{\omega_R}s\omega_2 &= P_{m2} - \omega_R D_2\omega_2 + \omega_R\left[D_{12} + \frac{K_{12}}{s}\right](\omega_1 - \omega_2) \\
&\quad - \omega_R\left[D_{23} + \frac{K_{23}}{s}\right](\omega_2 - \omega_3) \\
&\cdots \\
\frac{2W_{kn}}{\omega_R}s\omega_n &= P_{mn} - \omega_R D_n\omega_n + \omega_R\left[D_{mn} + \frac{K_{mn}}{s}\right](\omega_m - \omega_n) \\
&\quad - \omega_R\left[D_{ng} + \frac{K_{ng}}{s}\right](\omega_n - \omega_g) \\
\frac{2W_{kg}}{\omega_R}s\omega_g &= -P_e - \omega_R D_g\omega_g + \omega_R\left[D_{ng} + \frac{K_{ng}}{s}\right](\omega_n - \omega_g) \\
&\quad - \omega_R\left[D_{ga} + \frac{K_{ga}}{s}\right](\omega_g - \omega_a) \\
\frac{2W_{kq}}{\omega_R}s\omega_b &= 0 - \omega_{XR} D_q\omega_b + \omega_{XR}R_G\left[D_{ga} + \frac{K_{ga}}{s}\right](\omega_g - \omega_a) \\
&\quad - \omega_{XR}\left[D_{bx} + \frac{K_{bx}}{s}\right](\omega_b - \omega_x) \\
\frac{2W_{kx}}{\omega_R}s\omega_x &= -P_x - \omega_{XR} D_x\omega_x + \omega_{XR}\left[D_{bx} + \frac{K_{bx}}{s}\right](\omega_b - \omega_x).
\end{aligned} \tag{4.25}$$

We now introduce another change to the equations that will be useful in later work. The generator is, of course, connected to an electrical network, the equations for which were developed in Chapter 3. The network operates at some frequency that is always close to the rated system frequency or base frequency, which is given by

$$\text{Rated System Angular Velocity} = \omega_B = 2\pi f_B \tag{4.26}$$

where we use the subscript "B" to indicate a base electrical quantity. This angular frequency is related to the mechanical angular velocity of the synchronous generator by the number of poles in the generator.

$$\frac{p}{2} = \frac{\omega_B}{\omega_R} = \frac{\omega_e}{\omega_g} \tag{4.27}$$

where p is the number of poles. This relation is seen to hold whether expressed in terms of rated velocities or actual velocities. Equation (4.27) introduces a new term, the actual electrical angular frequency of the generator, which is often called simply "ω". We add the subscript "e" here for clarity since there are so many angular velocity terms in the equations. This electrical angular frequency is exactly the same as that defined for the generator in (2.1) and (3.2).

From (4.27) we derive the relationship

$$\omega_g = \frac{\omega_R}{\omega_B}\omega_e \quad \text{mechanical rad/ s} \tag{4.28}$$

which may be substituted into (4.25) to give the generator angular velocity always in terms of the electrical angular frequency. This couples the shaft equations to the generator and network equations in a more effective way. If this is done, the generator equation of (4.25) becomes

$$\begin{aligned}\frac{2W_{kg}}{\omega_B}s\omega_e = -P_e - \frac{\omega_R^2}{\omega_B}D_g\omega_e + \omega_R\left[D_{ng} + \frac{K_{ng}}{s}\right]\left(\omega_n - \frac{\omega_R}{\omega_B}\omega_e\right) \\ - \omega_R\left[D_{ga} + \frac{K_{ga}}{s}\right]\left(\frac{\omega_R}{\omega_B}\omega_e - \omega_a\right)\end{aligned} \tag{4.29}$$

A similar substitution can be made in the equations for the generator angular velocity for the gear system and also for the nth turbine.

4.4 NORMALIZATION OF THE SHAFT EQUATIONS

The purpose of this section is to normalize the shaft equations for incorporation into the total system analysis on a unified basis. Both the generators and the network elements are normalized to some base voltampere quantity, which is selected arbitrarily, and all system elements are expressed in a per unit of this base quantity. It is well known that two independent base quantities must be chosen to completely define the per unit system. The base voltampere quantity is almost universally chosen as one of these base values.

We now define the base quantity

$$S_{B3} = \text{Base 3 Phase Voltamperes.} \tag{4.30}$$

We assume here that this quantity is arbitrarily chosen, although many practitioners take this to be some convenient number such as 100 MVA.

The shaft power equations (4.25) are in mks units, and therefore are expressed in watts and have the dimensions of voltamperes. If we divide each of these equations by (4.30), the result is a dimensionless (per unit) quantity that is proportional to the base voltampere chosen at the outset. If we perform this operation, the following equations result:

$$\frac{2W_{k1}}{\omega_R S_{B3}} s\omega_1 = \frac{P_{m1}}{S_{B3}} - \frac{\omega_R D_1}{S_{B3}}\omega_1 - \frac{\omega_R}{S_{B3}}\left[D_{12} + \frac{K_{12}}{s}\right](\omega_1 - \omega_2)$$

$$\frac{2W_{k2}}{\omega_R S_{B3}} s\omega_2 = \frac{P_{m2}}{S_{B3}} - \frac{\omega_R D_2}{S_{B3}}\omega_2 + \frac{\omega_R}{S_{B3}}\left[D_{12} + \frac{K_{12}}{s}\right](\omega_1 - \omega_2)$$
$$- \frac{\omega_R}{S_{B3}}\left[D_{23} + \frac{K_{23}}{s}\right](\omega_2 - \omega_3)$$

$$\cdots$$

$$\frac{2W_{kn}}{\omega_R S_{B3}} s\omega_n = \frac{P_{mn}}{S_{B3}} - \frac{\omega_R D_n}{S_{B3}}\omega_n + \frac{\omega_R}{S_{B3}}\left[D_{mn} + \frac{K_{mn}}{s}\right](\omega_m - \omega_n)$$
$$- \frac{\omega_R}{S_{B3}}\left[D_{ng} + \frac{K_{ng}}{s}\right](\omega_n - \omega_g)$$

$$
\begin{aligned}
\frac{2W_{kg}}{\omega_R S_{B3}} s\omega_e &= \frac{-P_e}{S_{B3}} - \frac{\omega_R^2 D_g}{\omega_B S_{B3}}\omega_e + \frac{\omega_R}{S_{B3}}\left[D_{ng} + \frac{K_{ng}}{s}\right]\left(\omega_n - \frac{\omega_R}{\omega_B}\omega_g\right) \\
&\quad - \frac{\omega_R}{S_{B3}}\left[D_{ga} + \frac{K_{ga}}{s}\right]\left(\frac{\omega_R}{\omega_B}\omega_e - \omega_a\right) \\
\frac{2W_{kq}}{\omega_{XR} S_{B3}} s\omega_b &= 0 - \frac{\omega_{XR} D_q}{S_{B3}}\omega_b + \frac{\omega_{XR} R_G}{S_{B3}}\left[D_{ga} + \frac{K_{ga}}{s}\right]\left(\frac{\omega_R}{\omega_B}\omega_e - \omega_a\right) \\
&\quad - \frac{\omega_{XR}}{S_{B3}}\left[D_{bx} + \frac{K_{bx}}{s}\right](\omega_b - \omega_x) \\
\frac{2W_{kx}}{\omega_{XR} S_{B3}} s\omega_x &= \frac{-P_x}{S_{B3}} - \frac{\omega_{XR} D_x}{S_{B3}}\omega_x + \frac{\omega_{XR}}{S_{B3}}\left[D_{bx} + \frac{K_{bx}}{s}\right](\omega_b - \omega_x).
\end{aligned}
\tag{4.31}
$$

Now, from the definition of the gear ratio in terms of rated angular velocities (4.7) we may write

$$
\frac{\omega_{XR} R_G}{S_{B3}} = \frac{\omega_R}{S_{B3}} \tag{4.32}
$$

which can be used to simplify one of the coefficients in the equation for the equivalent gear box power.

We now define the normalized coefficients as follows:

$$
\begin{aligned}
H_i &= \frac{W_{ki}}{S_{B3}} \quad \text{seconds} \\
D_{ui} &= \frac{D_i \omega_{Ri}^2}{S_{B3}} \quad \text{per unit} \\
K_{ui} &= \frac{K_i \omega_{Ri}}{S_{B3}} \quad \text{sec}^{-1}
\end{aligned}
\tag{4.33}
$$

We also note that the per unit power is defined as

$$
P_{ui} = \frac{P_i}{S_{B3}} \quad \text{per unit} \tag{4.34}
$$

The inertia constant "H" is exactly as defined elsewhere [1] for studies involving the normalized modeling of turbine-generator shaft inertias. Note that, although a normalized quantity, the inertia constant is not dimensionless since it is the coefficient of a time derivative term in the differential equations. We use a notation that places the subscript "u" on these coefficients to signify a per unit quantity. Later, when all quantities are normalized, these "u" subscripts can be dropped.

Substituting (4.26) - (4.34) into (4.25) we write the equations

$$
\begin{aligned}
2H_1 s\frac{\omega_1}{\omega_R} &= P_{um1} - D_{u1}\frac{\omega_1}{\omega_R} - \left(D_{u12} + \frac{K_{u12}}{s}\right)\left(\frac{\omega_1 - \omega_2}{\omega_R}\right) \\
2H_2 s\frac{\omega_2}{\omega_R} &= P_{um2} - D_{u2}\frac{\omega_2}{\omega_R} + \left(D_{u12} + \frac{K_{u12}}{s}\right)\left(\frac{\omega_1 - \omega_2}{\omega_R}\right) \\
&\quad - \left(D_{u23} + \frac{K_{u23}}{s}\right)\left(\frac{\omega_2 - \omega_3}{\omega_R}\right) \\
&\cdots \\
2H_n s\frac{\omega_n}{\omega_R} &= P_{umn} - D_{un}\frac{\omega_n}{\omega_R} + \left(D_{umn} + \frac{K_{umn}}{s}\right)\left(\frac{\omega_m - \omega_n}{\omega_R}\right) \\
&\quad - \left(D_{ung} + \frac{K_{ung}}{s}\right)\left(\frac{\omega_n - \omega_g}{\omega_R}\right) \\
2H_g s\frac{\omega_e}{\omega_B} &= -P_{ue} - D_{ug}\frac{\omega_e}{\omega_B} + \left(D_{ung} + \frac{K_{ung}}{s}\right)\left(\frac{\omega_n}{\omega_R} - \frac{\omega_e}{\omega_B}\right) \\
&\quad - \left(D_{ugb} + \frac{K_{ugb}}{s}\right)\left(\frac{\omega_e}{\omega_B} - \frac{\omega_a}{\omega_R}\right) \\
2H_b s\frac{\omega_b}{\omega_{XR}} &= 0 - D_{ub}\frac{\omega_b}{\omega_{XR}} + \left(D_{ugb} + \frac{K_{ugb}}{s}\right)\left(\frac{\omega_e}{\omega_B} - \frac{\omega_a}{\omega_R}\right) \\
&\quad - \left(D_{ubx} + \frac{K_{ubx}}{s}\right)\left(\frac{\omega_b - \omega_x}{\omega_{XR}}\right) \\
2H_x s\frac{\omega_x}{\omega_{XR}} &= -P_{ux} - D_{ux}\frac{\omega_x}{\omega_{XR}} + \left(D_{ubx} + \frac{K_{ubx}}{s}\right)\left(\frac{\omega_b - \omega_x}{\omega_{XR}}\right).
\end{aligned}
\tag{4.35}
$$

Note that all quantities in the above equations are per unit quantities except the angular velocities, which are in radians per second. These velocities, however, are all divided by the base for that part of the shaft system. Hence, these *ratios* are per unit angular velocities, but referred to different base velocities that correspond to the origin of the measurement. Since all quantities are in per unit, the "u" subscripts may be neglected in the future writing of these same equations and the angular velocities may also be written in per unit. This will henceforth be done.

An interesting analogy can be drawn between the shaft model and the network model. In the network model, we normalize the equations by selecting one voltampere base for the entire network and a different base voltage for all branches and nodes at a given rated voltage, and with ideal transformers connecting the different voltages. In the shaft equations we again select one voltampere base, but we select different base angular velocities for each shaft section, with the shaft sections connected by ideal gear ratios. The two concepts are completely analagous.

4.5 THE INCREMENTAL SHAFT EQUATIONS

For eigenvalue computation, we use an incremental form of all equations. The generator equations are nonlinear and must be linearized about the initial operating point in an incremental fashion. The network equations are linear and are exactly the same equations whether in incremental or exact form. The normalized shaft power equations derived in the previous section are also linear and can be written immediately in incremental form by replacing each variable with an incremental variable. Thus we write

$$x = x_o + \Delta x \tag{4.36}$$

as the general statement of one of the system variables. The shaft equations given in (4.35) are in terms of the system variables, which we designate as "x". We replace each of these variables with Δx in the incremental form of the equations. This will make the equations immediately compatible with the linearized machine and network equations.

Before writing the incremental form, however, we make one additional change in the arrangement of the equations. We illustrate the procedure by taking one of the equations as an example, viz., the jth equation, with the angular velocity in per unit.

$$\begin{aligned} 2H_j s\,\Delta\omega_j = \Delta P_{mj} - D_j\,\Delta\omega_j &+ \left[D_{ij} + \frac{K_{ij}}{s}\right](\Delta\omega_i - \Delta\omega_j) \\ &- \left[D_{jk} + \frac{K_{jk}}{s}\right](\Delta\omega_j - \Delta\omega_k) \end{aligned} \tag{4.37}$$

This equation may be rewritten by rearranging the terms to obtain the following form.

$$(2H_j s + D_j)\Delta\omega_j = \Delta P_{mj} + \left[D_{ij} + \frac{K_{ij}}{s}\right](\Delta\omega_i - \Delta\omega_j) - \left[D_{jk} + \frac{K_{jk}}{s}\right](\Delta\omega_j - \Delta\omega_k) \tag{4.38}$$

or

$$\Delta\omega_j = \frac{1}{2H_j s + D_j}\left[\Delta P_{mj} + \left[D_{ij} + \frac{K_{ij}}{s}\right](\Delta\omega_i - \Delta\omega_j) - \left[D_{jk} + \frac{K_{jk}}{s}\right](\Delta\omega_j - \Delta\omega_k)\right] \tag{4.39}$$

Equation (4.39) gives an expression for the per unit incremental angular velocity of the jth turbine, computed in terms of the incremental turbine input power and the interactions with turbine sections adjacent to the jth turbine. Using this approach, we write the entire shaft system equations as follows.

$$\Delta\omega_1 = \frac{1}{2H_1 s + D_1}\left[\Delta P_{m1} - \left[D_{12} + \frac{K_{12}}{s}\right](\Delta\omega_1 - \Delta\omega_2)\right]$$

$$\Delta\omega_2 = \frac{1}{2H_2 s + D_2}\left[\Delta P_{m2} + \left(D_{12} + \frac{K_{12}}{s}\right)(\Delta\omega_1 - \Delta\omega_2) - \left(D_{23} + \frac{K_{23}}{s}\right)(\Delta\omega_2 - \Delta\omega_3)\right]$$

...

$$\Delta\omega_n = \frac{1}{2H_n s + D_n}\left[\Delta P_{mn} + \left(D_{mn} + \frac{K_{mn}}{s}\right)(\Delta\omega_m - \Delta\omega_n) - \left(D_{ng} + \frac{K_{ng}}{s}\right)(\Delta\omega_n - \Delta\omega_e)\right]$$

$$\Delta\omega_e = \frac{1}{2H_g s + D_g}\left[-\Delta P_e + \left(D_{ng} + \frac{K_{ng}}{s}\right)(\Delta\omega_n - \Delta\omega_e) - \left(D_{ga} + \frac{K_{ga}}{s}\right)(\Delta\omega_e - \Delta\omega_a)\right]$$

$$\Delta\omega_b = \frac{1}{2H_q\, s + D_q}\left[0 + \left(D_{ga} + \frac{K_{ga}}{s}\right)(\Delta\omega_e - \Delta\omega_a) - \left(D_{bx} + \frac{K_{bx}}{s}\right)(\Delta\omega_b - \Delta\omega_x)\right]$$

$$\Delta\omega_x = \frac{1}{2H_x\, s + D_x}\left[-\Delta P_x + \left[D_{bx} + \frac{K_{bx}}{s}\right](\Delta\omega_b - \Delta\omega_x)\right] \tag{4.40}$$

The above equations are in the desired form for eigenvalue analysis. Note that all angular velocities in the above equation are in per unit based on the rated angular velocity of that part of the shaft.

4.6 THE TURBINE MODEL

This section presents a model of the individual turbine cylinders that produces the input power for the various sections of the shaft model. Turbine models differ according to the physical structure and arrangement of the energy supply system. A general turbine model can be developed so that it can be used for nearly all arrangements of either steam or hydraulic turbines. Such a model was introduced in an IEEE Committee report [3], and is the model that will be used here.

The general turbine model is shown in Figure 4.3. This model will be discussed in terms of its application to steam turbine modeling, since this is of greatest interest in SSR analysis, but the model is equally applicable to hydro turbine modeling.

A major difference in steam turbine-generator systems is associated with the grouping of turbines and generators on one or more shafts. Almost all turbine systems are compounded, i.e., more than one turbine cylinder is required to produce full output power. Two major classifications of compound units are tandem-compound and cross-compound units. Tandem-compound units have all turbine cylinders on the same shaft and large units may have six to ten turbines mounted on the same shaft in addition to the generator and possibly the exciter. Cross-compound turbine systems divide the turbine cylinders between two shafts, each of which will also have its own generator and exciter. Cross-compound units have the advantage that the shafts are much shorter for the same rating. The disadvantage is that two generators are required for cross-compound units.

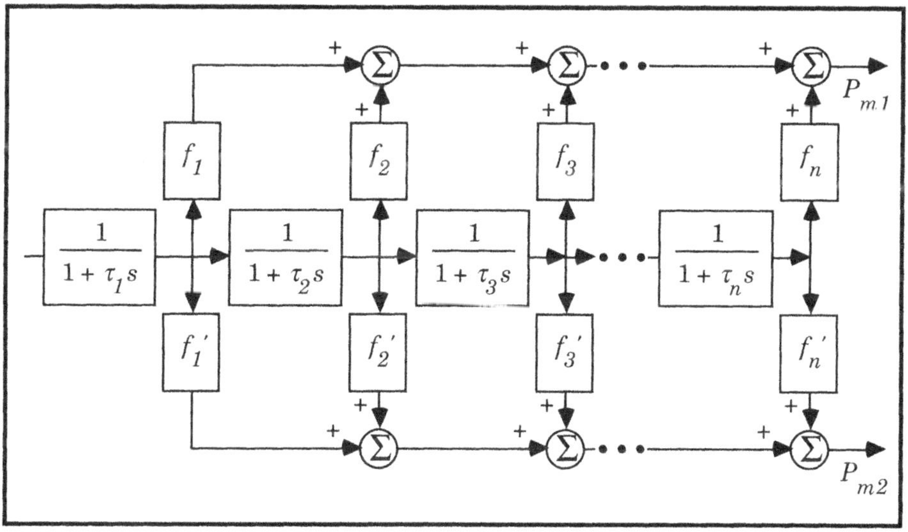

Figure 4.3 General Block Diagram of a Turbine Transfer Function

The shaft equations derived herein apply to either cross-compound or tandem-compound shafts.

Tandem-compound turbines are modeled using only the top half of the block diagram of Figure 4.3, resulting in power output P_{m1}. Cross-compound units require the entire block diagram with the power output divided between the two shafts. The time constants are, from left to right, the steam bowl, the first reheat time lag, the second reheat time lag or low pressure crossover lag, etc. Two reheats are usually the maximum in practice. The fractions "f" are the fractions of the total power output produced by the several turbines. These fractions must sum to unity for each turbine shaft.

There are several parts of the steam supply system that are important in the dynamic model of the turbine system because of the size of the time lag associated with the steam piping and storage volumes. These are approximated in most models by treating these volumes as first order lags. Some of these lags are significant, particularly that due to the reheater volumes. These important lags are represented in the turbine model such that there is a one-to-one correspondence between the block diagram and the physical system structure. This simplifies the problem of estimating the time constants based on the physical dimensions and volumes of the actual steam supply system. An example showing the model of a tandem-

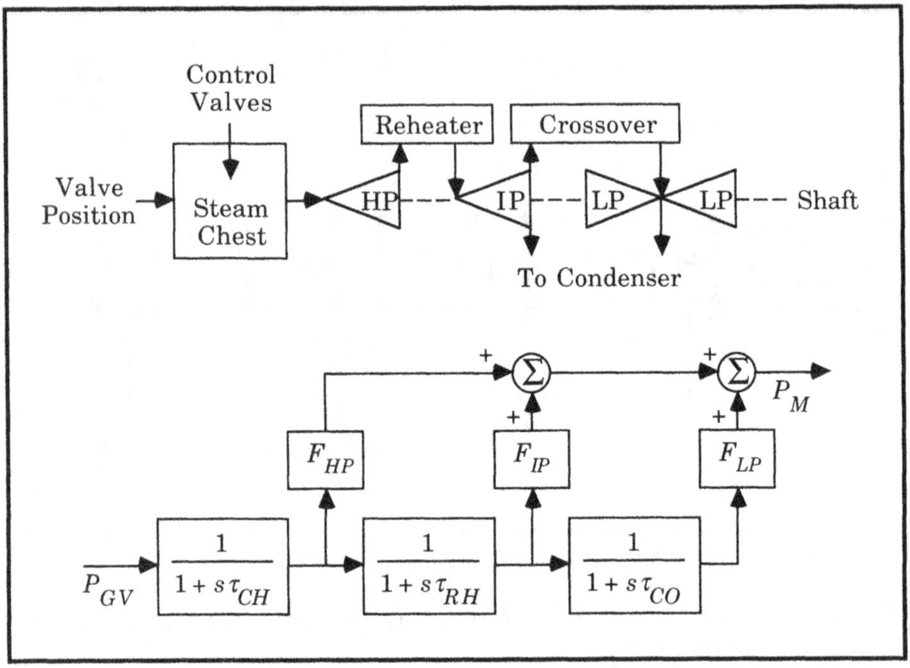

Figure 4.4 Mathematical Model of a Tandem-Compound, Single-Reheat Steam Turbine

compound, single-reheat unit is shown in Figure 4.4, where the upper part of the figure shows the physical arrangements of turbine cylinders and the lower part shows the mathematical model, which should be recognized as utilizing a portion of the general model structure of Figure 4.3.

4.7 THE COMPLETE TURBINE AND SHAFT MODEL

We now combine the incremental shaft model of Section 4.5 with an incremental turbine model to develop a block diagram of the entire process that describes a given turbine shaft system. This block diagram is shown in Figure 4.5.

Referring to Figure 4.5 we recognize the two major segments of the turbine shaft description. The turbine model is that portion to the upper left and is exactly the same as Figure 4.3, except that only one shaft is shown. The right portion of Figure 4.5 is a graphical representation of equation (4.37). The inputs to the turbine shaft system is an incremental change in the

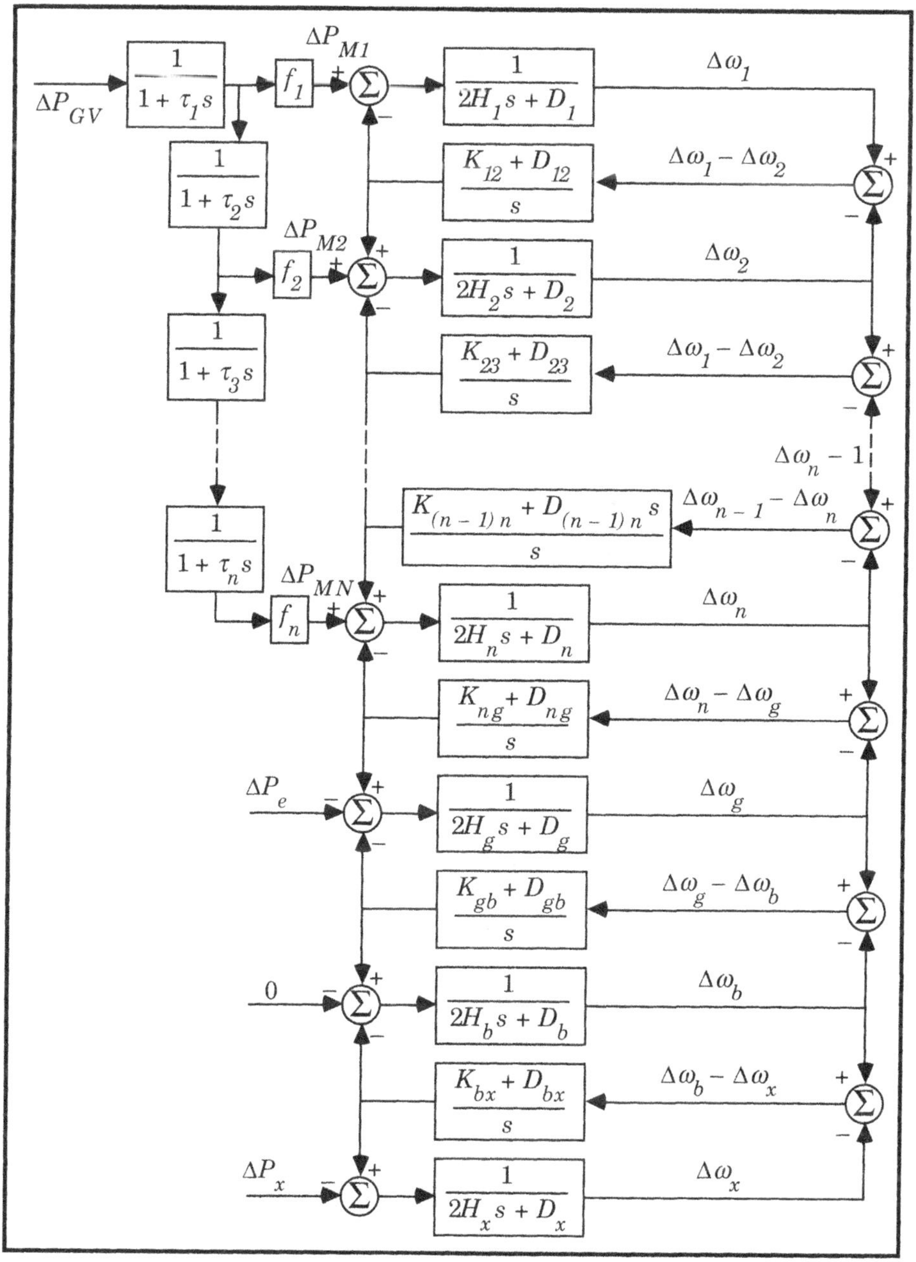

Figure 4.5 Block Diagram of the Turbine Shaft System

governor valve position P_{GV} (see Figure 1.3). The coupling to the generator system equations is through the electromagnetic power term ΔP_e. The coupling to the excitation system, if the exciter is represented, is through the exciter electromagnetic power ΔP_x. Both of these electromagnetic power terms enter the diagram with a negative sign since they constitute mechanical loads to the system and represent decelerating torques in the Newton equations.

The engineer could add a governor model, the output of which will be the valve position of the control valves, which is the input to the turbine model. This addition is straightforward and will not be presented formally, since it is not really required for the purpose at hand, which is the identification of any resonances between turbine-generator shafts and network modes of oscillation. In many cases, the analysis for SSR would omit the turbine model altogether. The turbine model has very large reheat time constants that greatly effect the dynamic response of the generating unit, but these large time constants have little effect on the natural frequencies of oscillation and the coupling between the machine and the network.

Example 4.1 Extend Example 3.2 to add the simple shaft model shown in Figure 4.6, where we have assumed two steam turbines and one generator on the shaft. For this simple shaft model, we write the following differential equations:

$$
\begin{aligned}
\dot{\omega}_1 &= \frac{1}{2H_1}P_{m1} - \frac{D_1}{2H_1}\omega_1 - \frac{K_{12}}{2H_1}(\theta_1 - \theta_2) \\
\dot{\omega}_2 &= \frac{1}{2H_2}P_{m2} - \frac{D_2}{2H_2}\omega_2 + \frac{K_{12}}{2H_2}(\theta_1 - \theta_2) - \frac{K_{2g}}{2H_2}(\theta_2 - \theta_g) \\
\dot{\omega} &= \frac{-1}{2H_g}P_e - \frac{D_g}{2H_g}\omega + \frac{K_{2g}}{2H_g}(\theta - \theta_2)
\end{aligned}
\tag{4.41}
$$

$$
\begin{aligned}
\dot{\theta}_1 &= \omega_1 \\
\dot{\theta}_2 &= \omega_2 \\
\dot{\theta} &= \omega
\end{aligned}
\tag{4.42}
$$

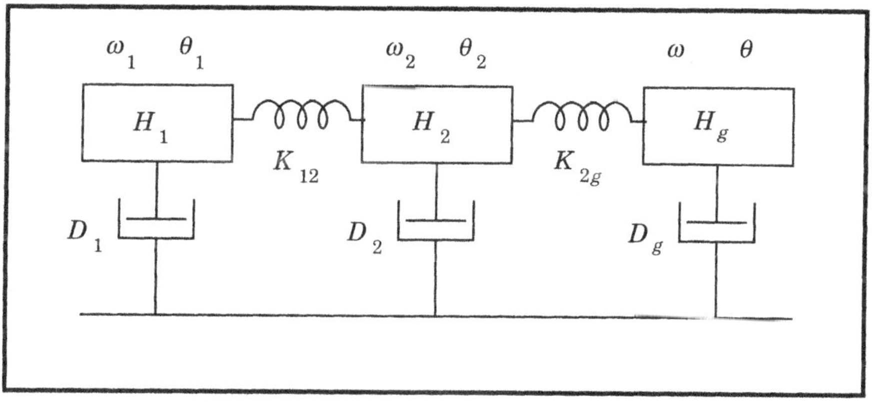

Figure 4.6 Shaft Model for Example 4.1

where

ω_1 = speed of mass 1
ω_2 = speed of mass 2
ω = speed of generator
θ_1 = angle of mass 1
θ_2 = angle of mass 2
θ = angle of generator mass

These six equations replace the speed equation of (3.61), with the result shown in (4.43). The matrix elements are as follows:

$$A_{7-7} = \frac{-D_1}{2H_1} \qquad A_{7-11} = \frac{+K_{12}}{2H_1} \qquad A_{9-11} = \frac{+K_{2g}}{2H_2} \qquad A_{9-3} = \frac{I_{dD}}{2H_g}$$

$$A_{8-8} = \frac{-D_2}{2H_2} \qquad A_{8-10} = \frac{-K_{12}}{2H_2} \qquad A_{9-12} = \frac{-K_{1g}}{2H_2} \qquad A_{9-4} = \frac{-I_{qq}}{2H_g}$$

$$A_{9-9} = \frac{-D_g}{2H_g} \qquad A_{8-11} = \frac{K_{12} - K_{2g}}{2H_2} \qquad A_{9-1} = \frac{I_{dd}}{2H_g} \qquad A_{9-5} = \frac{-I_{qG}}{2H_g}$$

$$A_{7-10} = \frac{-K_{12}}{2H_1} \qquad A_{8-12} = \frac{+K_{2g}}{2H_2} \qquad A_{9-2} = \frac{I_{dF}}{2H_g} \qquad A_{9-6} = \frac{-I_{qQ}}{2H_g}$$

$$A_{13-13} = -\frac{1}{\tau_R}$$

$$A_{16-1} = S\left(\Gamma_{dd}h_{dd} + \Gamma_{dF}h_{Fd} + \Gamma_{dD}h_{Dd}\right)$$

$$A_{14-13} = \frac{-K_F K_A}{\tau_F \tau_A}$$

$$A_{17-1} = S\left(\Gamma_{qq}\omega_B\omega_o\right)$$

$$A_{16-2} = S\left(\Gamma_{dd}h_{dF} + \Gamma_{dF}h_{FF} + \Gamma_{dD}h_{DF}\right)$$

$$A_{15-13} = \frac{-K_A}{\tau_A}$$

$$A_{16-3} = S\left(\Gamma_{dd}h_{dD} + \Gamma_{dF}h_{FD} + \Gamma_{dD}h_{DD}\right)$$

$$A_{16-4} = -S\Gamma_{dd}\omega_B\omega_o$$
$$A_{17-4} = S\left(\Gamma_{qq}h_{qq} + \Gamma_{qG}h_{Gq} + \Gamma_{qQ}h_{Qq}\right)$$
$$A_{17-5} = S\left(\Gamma_{qq}h_{qG} + \Gamma_{qG}h_{GG} + \Gamma_{qQ}h_{QG}\right)$$
$$A_{17-6} = S\left(\Gamma_{qq}h_{qQ} + \Gamma_{qG}h_{GQ} + \Gamma_{qQ}h_{QQ}\right)$$
$$A_{16-9} = -S\Gamma_{dd}\omega_B\psi_{qo} + v_{qo}$$
$$A_{17-9} = +S\Gamma_{qq}\omega_B\psi_{do} - v_{do}$$
$$A_{13-16} = \frac{v_{do}K_R}{V_{to}\tau_R}$$
$$B_{14-3} = \frac{K_F K_A}{\tau_F \tau_A}$$
$$B_{14-4} = \frac{K_F K_A}{\tau_F \tau_A}$$

$$A_{14-14} = \frac{-1}{\tau_F}\left(\frac{K_F K_A}{\tau_A} + 1\right)$$
$$A_{15-14} = A_{15-13}$$
$$A_{15-15} = \frac{-1}{\tau_A}$$
$$A_{16-15} = S\Gamma_{dF}k_x$$
$$A_{16-16} = -S\Gamma_{dd}\omega_B$$
$$A_{13-17} = \frac{v_{qo}K_R}{V_{to}\tau_R}$$
$$A_{17-17} = -S\Gamma_{qq}\omega_B$$
$$B_{15-3} = \frac{K_A}{\tau_A}$$
$$B_{15-4} = \frac{K_A}{\tau_A}$$

The three-mass shaft model is a sixth order system of equations, which replaces the single angular velocity equation of the previous examples. This set of equations could be further expanded to include a speed governor, which would define the mechanical power inputs in terms of additional state variables. This additional detail is not required, however, if the goal is simply to determine the existence of torsional interaction in the system under study. This completes Example 4.1.

The previous example illustrates the modeling detail that is required for the analysis of SSR to determine torsional interactions between the turbine-generator and the electrical system. This technique can be extended to much larger systems that include several machines and networks of many nodes and branches. Obviously, this would be difficult to do manually as we have done here. Computer programs are available that have been prepared especially for type of analysis, so that the user needs only to specify the data for the generators, shafts, exciters, and network. Even the specification of the required data is difficult, due to the uncertainty or absence of certain parameters. The problem of determining accurate data will be explored more fully in Part 3 of the book.

$$
\begin{bmatrix} \dot{\psi}_d \\ \dot{\psi}_F \\ \dot{\psi}_D \\ \hline \dot{\psi}_q \\ \dot{\psi}_G \\ \dot{\psi}_Q \\ \hline \dot{\omega}_1 \\ \dot{\omega}_2 \\ \dot{\omega} \\ \hline \dot{\theta}_1 \\ \dot{\theta}_2 \\ \dot{\theta} \\ \hline \dot{V}_1 \\ \dot{V}_2 \\ \dot{E}_{FD} \\ \hline \dot{v}_d \\ \dot{v}_q \end{bmatrix}
=
\left[\begin{array}{ccc|ccc|ccc|ccc|ccc|cc}
h_{dd} & h_{dF} & h_{dD} & -\omega_B\omega_o & 0 & 0 & 0 & 0 & -\omega_B\psi_{qo} & 0 & 0 & 0 & 0 & 0 & 0 & -\omega_B & 0 \\
h_{Fd} & h_{FF} & h_{FD} & 0 & 0 & 0 & 0 & 0 & 0 & 0 & 0 & 0 & 0 & 0 & k_x & 0 & 0 \\
h_{Dd} & h_{DF} & h_{DD} & 0 & 0 & 0 & 0 & 0 & 0 & 0 & 0 & 0 & 0 & 0 & 0 & 0 & 0 \\
\hline
\omega_B\omega_o & 0 & 0 & h_{qq} & h_{qG} & h_{qQ} & 0 & 0 & \omega_B\psi_{do} & 0 & 0 & 0 & 0 & 0 & 0 & 0 & -\omega_B \\
0 & 0 & 0 & h_{Gq} & h_{GG} & h_{GQ} & 0 & 0 & 0 & 0 & 0 & 0 & 0 & 0 & 0 & 0 & 0 \\
0 & 0 & 0 & h_{Qq} & h_{QG} & h_{QQ} & 0 & 0 & 0 & 0 & 0 & 0 & 0 & 0 & 0 & 0 & 0 \\
\hline
0 & 0 & 0 & 0 & 0 & 0 & A_{7-7} & 0 & 0 & A_{7-10} & A_{7-11} & 0 & 0 & 0 & 0 & 0 & 0 \\
0 & 0 & 0 & 0 & 0 & 0 & 0 & A_{8-8} & 0 & A_{8-10} & A_{8-11} & A_{8-12} & 0 & 0 & 0 & 0 & 0 \\
A_{9-1} & A_{9-2} & A_{9-3} & A_{9-4} & A_{9-5} & A_{9-6} & 0 & 0 & A_{9-9} & 0 & A_{9-11} & A_{9-12} & 0 & 0 & 0 & 0 & 0 \\
\hline
0 & 0 & 0 & 0 & 0 & 0 & 1 & 0 & 0 & 0 & 0 & 0 & 0 & 0 & 0 & 0 & 0 \\
0 & 0 & 0 & 0 & 0 & 0 & 0 & 1 & 0 & 0 & 0 & 0 & 0 & 0 & 0 & 0 & 0 \\
0 & 0 & 0 & 0 & 0 & 0 & 0 & 0 & 1 & 0 & 0 & 0 & 0 & 0 & 0 & 0 & 0 \\
\hline
0 & 0 & 0 & 0 & 0 & 0 & 0 & 0 & 0 & 0 & 0 & 0 & A_{13-13} & 0 & 0 & A_{13-16} & A_{13-17} \\
0 & 0 & 0 & 0 & 0 & 0 & 0 & 0 & 0 & 0 & 0 & 0 & A_{14-13} & A_{14-14} & 0 & 0 & 0 \\
0 & 0 & 0 & 0 & 0 & 0 & 0 & 0 & 0 & 0 & 0 & 0 & A_{15-13} & A_{15-14} & A_{15-15} & 0 & 0 \\
\hline
A_{16-1} & A_{16-2} & A_{16-3} & A_{16-4} & 0 & 0 & 0 & 0 & A_{16-9} & 0 & 0 & 0 & 0 & 0 & A_{16-15} & A_{16-16} & \omega_B \\
A_{17-1} & 0 & 0 & A_{17-4} & A_{17-5} & A_{17-6} & 0 & 0 & A_{17-9} & 0 & 0 & 0 & 0 & 0 & 0 & -\omega_B & A_{17-17}
\end{array}\right]
\begin{bmatrix} \psi_d \\ \psi_F \\ \psi_D \\ \hline \psi_q \\ \psi_G \\ \psi_d \\ \hline \omega_1 \\ \omega_2 \\ \omega \\ \hline \theta_1 \\ \theta_2 \\ \theta \\ \hline V_1 \\ V_2 \\ E_{FD} \\ \hline v_d \\ v_q \end{bmatrix}
$$

$$
+ \left[\begin{array}{cccc}
0 & 0 & 0 & 0 \\
0 & 0 & 0 & 0 \\
0 & 0 & 0 & 0 \\
\hline
0 & 0 & 0 & 0 \\
0 & 0 & 0 & 0 \\
0 & 0 & 0 & 0 \\
\hline
1/2H_1 & 0 & 0 & 0 \\
0 & 1/2H_2 & 0 & 0 \\
0 & 0 & 0 & 0 \\
\hline
0 & 0 & 0 & 0 \\
0 & 0 & 0 & 0 \\
0 & 0 & 0 & 0 \\
\hline
0 & 0 & 0 & 0 \\
0 & 0 & B_{13-3} & B_{13-4} \\
0 & 0 & B_{14-3} & B_{14-4} \\
\hline
0 & 0 & 0 & 0 \\
0 & 0 & 0 & 0
\end{array}\right]
\begin{bmatrix} P_{m1} \\ P_{m2} \\ V_{REF} \\ V_S \end{bmatrix}
\qquad (4.43)
$$

4.8 REFERENCES FOR CHAPTER 4

1. Anderson, P. M. and A. A. Fouad, **Power System Control and Stability**, Iowa State University Press, 1976.

2. Bowler, C.E.J., Private Communication, General Electric Co., Schenectady, New York, December 1984.

3. IEEE Committee Report, "Dynamic Models for Steam and Hydro Turbines in Power System Studies," **IEEE Trans on Power Apparatus and Systems,** v PAS-92, n 6, Nov/Dec 1973, p 1904-1915.

Part 3

System Parameters

Synchronous Generator Model Parameters

Turbine-Generator Shaft Model Parameters

CHAPTER 5

SYNCHRONOUS GENERATOR MODEL PARAMETERS

This chapter and Chapter 6 discuss the problems encountered in determining of accurate parameters for the synchronous machine and turbine-generator shaft models. Usually, there is no problem in determining model parameters for the network elements. The generator parameters are available from the manufacturers, but these parameters are usually intended for transient stability studies, not for SSR. This means that the model parameters are inherently limited to about a 0-5 Hz bandwidth, which is of questionable adequacy for SSR studies. There are also problems with manufacturer's estimates of the turbine-generator shaft model parameters, many of which must be estimated heuristically. These problems are discussed in Chapter 6 and solutions are suggested for improving the shaft model parameter estimations.

The theory developed in Chapter 2 leads to equation (2.88), which expresses the flux linkages in terms of the self and mutual inductances of the several windings in the machine, and (2.89), which gives the machine voltages in terms of these inductances and associated resistances. These equations may be rearranged in terms of mutual and leakage inductances to give the equivalent circuit arrangement of Figure 5.1 for the passive elements of the synchronous generator. The data supplied that one can obtain for SSR studies is usually in one of the following forms:

1. Conventional stability format data from the manufacturer.
2. Measured frequency response data from field tests.
3. Calculated R and X vs Frequency data from the manufacturer.
4. Direct winding data for the equivalent circuit model of Figure 5.1 (or other model) from the manufacturer or derived from field tests.

The parameter identification process for each of the above is discussed below. We begin by reviewing the kind of data that is ordinarily provided by the generator manufacturers. This is followed by a discussion of field testing for improvement of the machine representation. Examples are given of actual field tests and the improvement achieved in the machine modeling as the result of testing.

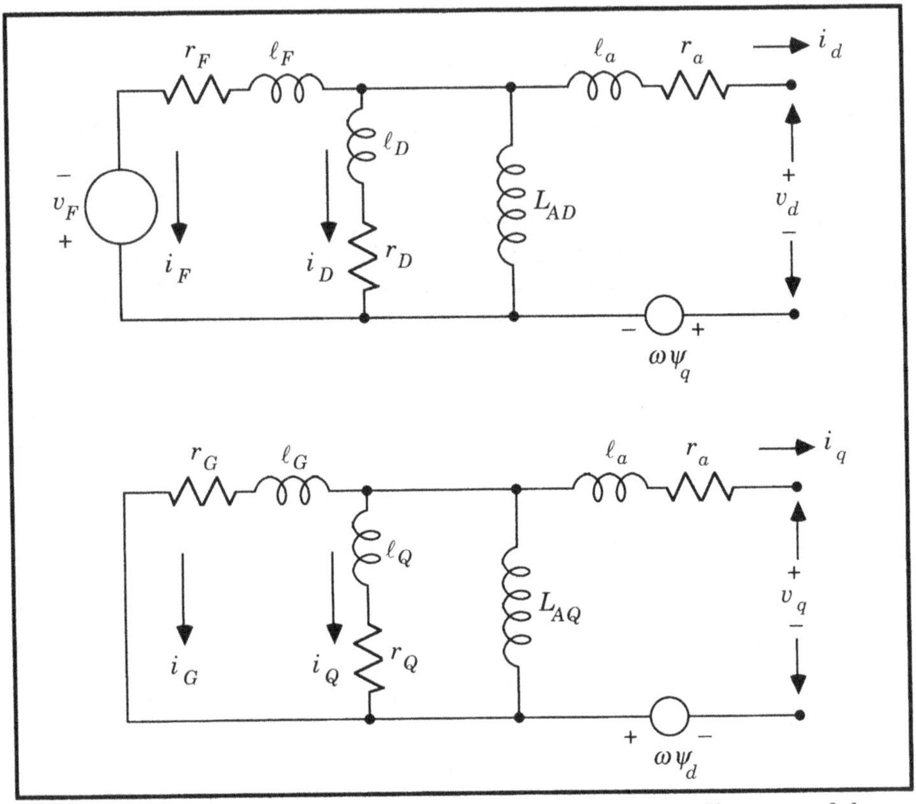

Figure 5.1 Two-axis Equivalent Circuits for the Passive Elements of the Synchronous Generator

5.1 CONVENTIONAL STABILITY DATA

Conventional stability format data supplied by the generator manufacturer is the most commonly available form of data. This data form was developed more than 50 years ago to provide information on the short circuit performance of the generator for circuit breaker and protective relay application [1]. Subsequently, these short circuit constants were extended for the purpose of calculating transient stability power limits. Today, these constants are widely used for all kinds of transient and dynamic studies.

In recent years the inadequacy of the conventional stability constants has been recognized and extensive efforts have been launched to obtain better parameters and/or better models. The need for this work on stability study constants has been championed largely by an IEEE Working Group formed for this purpose. Their work is delineated in part by a Working Group paper published in 1980 [2] and is discussed in detail in a record of their 1983 IEEE Symposium [3]. More recently, the Working Group, under the leadership of Paul Dandeno, has been developing standards for synchronous machines that address some of the problems in obtaining adequate data for simulation studies [4,5].

Based on field tests and correlation efforts of recent years, there is no reason to believe that the conventional stability data is entirely suitable for SSR studies, which deal mostly with rotor frequencies in the 10 to 50 Hz range. Even so, this being the most commonly available data, it is widely used for SSR studies. Recently, some manufacturers have supplied modified conventional stability data for use in SSR studies.

The parameter identification process using conventional stability data is performed as follows. A typical set of conventional stability data, their relationship to the symbols used in this report, and their typical values are shown in Table 5.1. The values quoted have been taken from the IEEE Second Benchmark Model [6].

The relationship between the parameters of the equivalent circuit of Figure 5.1 and the stability constants of Table 5.1 are given by (2.95) and the equations of Table 2.3. These relationships are repeated here in a form more suitable for calculations. Note that all parameters in (5.1) to (5.5) are in per unit except for ω_B, which is in radians per second.

A similar relationship is obtained for the q axis by replacing the subscripts appropriately.

It should be noted that (5.1) through (5.5) can be easily solved for the rotor circuit parameters if one starts from (5.1) and solves (5.2) through (5.5) in sequence. Each equation has only one unknown element, if the work is performed in this way.

$$L_d = L_{AD} + \ell_a \qquad \text{per unit} \qquad (5.1)$$

$$L_d' = \ell_a + \frac{1}{\dfrac{1}{L_{AD}} + \dfrac{1}{\ell_F}} \qquad \text{per unit} \qquad (5.2)$$

Table 5.1 Typical Values of Conventional Stability Data [6]

Symbol: Conventional Stability	Symbol: Used in This Report	Name of Parameter	Typical Per Unit Value
Reactances	Inductances		
X_L	ℓ_d ℓ_q ℓ_a	Armature Leakage	0.14
X_d	L_d	Direct Axis Synchronous	1.65
X_q	L_q	Quadrature Axis Synchronous	1.59
X_d'	L_d'	Direct Axis Transient	0.25
X_q'	L_q'	Quadrature Axis Transient	0.46
X_d''	L_d''	Direct Axis Subtransient	0.20
X_q''	L_q''	Quadrature Axis Subtransient	0.20
X_0	L_0	Zero Sequence	0.20
X_2	L_2	Negative Sequence	0.20
Time Constants			
T_{do}'	τ_{do}'	Direct Axis Open Circuit Transient	4.50
T_{qo}'	τ_{qo}'	Quadrature Axis Open Circuit Transient	0.55
T_{do}''	τ_{do}''	Direct Axis Open Circuit Subtransient	0.040
T_{qo}''	τ_{qo}''	Quadrature Axis Open Circuit Subtransient	0.090

$$L_d'' = \ell_a + \frac{1}{\dfrac{1}{L_{AD}} + \dfrac{1}{\ell_F} + \dfrac{1}{\ell_D}} \quad \text{per unit} \tag{5.3}$$

$$\tau_{do}' = \frac{L_{AD} + \ell_F}{\omega_B r_F} \quad \text{seconds} \tag{5.4}$$

$$\tau''_{do} = \frac{1}{\omega_B r_D}\left(\ell_D + \frac{1}{\frac{1}{L_{AD}} + \frac{1}{\ell_F}}\right) \quad \text{seconds} \qquad (5.5)$$

5.1.1 Approximations Involved In Parameter Computation

Equations (5.1)-(5.5) inherently included the following two assumptions [7]:

- The open circuit transient time constant of (5.4) is assumed to be due to the field circuit. This is equivalent to assuming no coupling effect of the damper winding, i.e., the damper resistance is infinite.
- The open circuit subtransient time constant definition (5.5) assumes zero resistance in the field winding.

The time constants calculated from (5.4) and (5.5) would be close to the time constant of the equivalent circuit under the assumption

$$r_D >> r_F.$$

The actual open circuit response of the flux for the machine represented by the equivalent circuit of Figure 5.1 will be characterized by two time constants that may differ significantly from the two time constants computed from (5.4) and (5.5), which are based on conventional stability data.

More exact procedures are available to fit the stability parameters to the equivalent circuit of Figure 5.1. One such approach is given in [8]. This technique involves an iterative solution of 14 nonlinear equations.

The authors are of the opinion that the more exact techniques are seldom justified if the conventional stability data is used. This is because the error introduced due to use of conventional stability data is far greater than the error due to approximations in the parameter identification process using (5.1) to (5.5). If, however, the conventional stability data has been modified to be suitable for use in SSR studies, then a more exacting fit, such as that given in [8], may be worthwhile.

5.2 MEASURED DATA FROM FIELD TESTS

Frequency response methods have been used to derive the transfer functions describing the response of machine flux to stator current and field voltage disturbances. Many approaches have been considered in an effort to obtain improved machine electrical constants through the application of frequency response methods. To date the major emphasis of the frequency response measurements have been to obtain better data in the lower frequency range for stability study purposes. Even so, the same methods can be used to obtain improved frequency response data in the frequency range of interest for SSR studies. However, to the authors' knowledge, this technique has not been pursued for obtaining SSR circuit models.

The development of an equivalent circuit model from such tests requires that the test data be presented in the form of operational inductances and/or transfer functions. For example, the measured direct-axis operational inductance might be approximated with the following second order transfer function.

$$L_d(s) = L_d \frac{(1+\tau_1 s)(1+\tau_2 s)}{(1+\tau_3 s)(1+\tau_4 s)}. \tag{5.6}$$

The time constants in (5.6) can then be related to the two rotor circuit model, such as that of Figure 5.1, to obtain the parameters of the model. The number of rotor circuits in the model is equal to the order of the polynomials used to match the measured data.

Occasionally, the last step in obtaining the parameters of the equivalent circuit can be avoided and the measurement of operational impedances and transfer functions in the form of (5.6) can be directly used in the eigenvalue program. For example, (2.114) and (2.118) have the information needed for calculation of eigenvalues. The operational impedance and transfer function defined in these equations can be directly substituted from the approximate fits obtained from the measured data. The equivalent circuit used for the direct and quadrature axes are important since these circuits and the circuit parameters used in their construction determine the behavior of the synchronous machine. These inductance transfer functions become an essential part of the basic system equations, which are restated here as (5.7):

$$\begin{aligned} \psi_d(s) &= L_d(s)i_d + G_d(s)v_F \\ \psi_q(s) &= L_q(s)i_q + G_q(s)v_G. \end{aligned} \tag{5.7}$$

This equation is independent of the type of equivalent circuit used or the number of rotor circuits used. The measured data can be approximated by any order of polynomial and directly substituted into (5.7). This will allow the measured data to be approximated by the higher order polynomials and will result in better accuracy in the frequency range of interest for SSR studies.

The relationship among the operational impedances of (5.6), the equivalent circuit model of Figure 5.1, and the measured data are explained further by the following example.

Example 5.1 Operational Inductance Calculations

We examine the operational inductances for the Ontario Hydro synchronous machine at Lambton. The standard model for the Lambton generator is given in Table 5.2. This table provides the data for the generator model of Figure 5.1. (There is a slight difference in the notation used in Tables 2.3 and 5.2, with this difference resolved below). The per unit values given are given in Table 5.2.

Table 5.2 Manufacturer's Standard Stability Data for Ontario Hydro's Lambton Generator [11]	
$\ell_a = \ell_d = \ell_q = 0.160$	
$L_{AD} = 1.810$	$L_{AQ} = 1.707$
$\ell_D = L_{kd1} = 0.1737$	$\ell_q = L_{kq1} = 0.0638$
$r_D = R_{kd1} = 0.0109$	$r_Q = R_{kq1} = 0.0164$
$\ell_F = L_{fd} = 0.1171$	$\ell_G = L_{kq2} = 0.3833$
$r_F = R_{fd} = 0.001189$	$r_G = R_{kq2} = 0.0099$

The parameters of (5.6) are related to the model of Figure 5.1 by (5.1) - (5.5), with similar relationships for the q axis. These relationships are derived in Chapter 2 with the result

$$L_d(s) = \frac{L_d(1+\tau'_d s)(1+\tau''_d s)}{(1+\tau'_{do} s)(1+\tau''_{do} s)} \tag{5.8}$$

$$L_q(s) = \frac{L_q\left(1+\tau'_q s\right)\left(1+\tau''_q s\right)}{\left(1+\tau'_{qo} s\right)\left(1+\tau''_{qo} s\right)}.$$

From (5.1) to (5.5) we compute the following inductances in per unit.

$$L_d = L_{AD} + \ell_a = 1.810 + 0.160 = 1.970$$

$$L'_d = \ell_a + \frac{1}{\dfrac{1}{L_{AD}} + \dfrac{1}{\ell_F}} = 0.160 + \frac{1}{\dfrac{1}{1.81} + \dfrac{1}{0.1171}} = 0.270$$

We may also compute the following time constants in seconds.

$$\tau'_{do} = \frac{L_{AD} + \ell_F}{\omega_B r_F} = \frac{1.81 + 0.1171}{(377)(0.001189)} = 4.30 \text{ s}$$

$$\tau''_{do} = \frac{1}{\omega_B r_D}\left(\ell_D + \frac{1}{\dfrac{1}{L_{AD}} + \dfrac{1}{\ell_F}}\right)$$

$$= \frac{1}{(377)(0.0109)}\left(0.01737 + \frac{1}{\dfrac{1}{1.81} + \dfrac{1}{0.1171}}\right) = 0.0310 \text{ s}$$

$$\tau'_d = \tau'_{do}\frac{L'_d}{L_d} = \frac{(4.30)(0.270)}{1.97} = 0.589 \text{ s}$$

$$\tau''_d = \tau''_{do}\frac{L''_d}{L'_d} = \frac{(0.0310)(0.175)}{0.270} = 0.0201 \text{ s.}$$

Hence, (5.6) for the d axis is given by

$$L_d(s) = 1.97\frac{(1+0.589s)(1+0.0201s)}{(1+4.300s)(1+0.0310s)} = M_d\angle\theta_d.$$

where we have defined the magnitude M_d and the phase θ_d as noted in Table 5.3.

Table 5.3 Magnitude and Phase of the Operational Inductance in the d axis as a Function of Frequency in Hertz

Frequency in Hz	Magnitude M_d in pu	Magnitude M_d in dB	Phase of M_d θ_d in degrees
0.001	1.969	5.89	-1.31
0.002	1.967	5.88	-2.62
0.005	1.952	5.82	-6.53
0.010	1.903	5.59	-12.77
0.020	1.739	4.81	-23.68
0.040	1.350	2.60	-38.00
0.050	1.190	1.51	-42.35
0.100	0.729	-2.74	-48.76
0.200	0.447	-7.00	-42.93
0.500	0.304	-10.31	-25.53
1.000	0.276	-11.17	-16.48
2.000	0.261	-11.64	-13.46
5.000	0.228	-12.80	-14.34
10.000	0.198	-14.03	-12.27
20.000	0.182	-14.78	-7.72
50.000	0.176	-15.07	-3.32
100.000	0.175	-15.12	-1.68
200.000	0.175	-15.14	-0.85
500.000	0.175	-15.14	-0.33
1000.000	0.175	-15.14	-0.16

In exactly the same manner we may obtain the operational inductance for the q axis.

$$L_q(s) = 1.867\frac{(1+0.1419s)(1+0.0280s)}{(1+0.5601s)(1+0.0609s)} = M_q\angle\theta_q.$$

where we have defined the magnitude M_q and the phase θ_q as given in Table 5.4.

Tables 5.3 and 5.4 give the magnitude and phase of these transfer functions as a function of frequency in hertz. The magnitude is given in both per unit and in decibels (dB), and the phase angle is given in degrees.

Table 5.4 Magnitude and Phase of the Operational Inductance in the q axis as a Function of Frequency in Hertz

Frequency in Hz	Magnitude M_q in pu	Magnitude M_q in dB	Phase of M_q θ_q in degrees
0.001	1.867	5.42	-0.16
0.002	1.867	5.42	-0.31
0.005	1.867	5.42	-0.79
0.010	1.866	5.42	-1.58
0.020	1.862	5.40	-3.16
0.040	------	-----	------
0.050	1.840	5.30	-7.87
0.100	1.767	4.94	-15.15
0.200	1.550	3.80	-26.78
0.500	0.995	-0.05	-41.32
1.000	0.648	-3.76	-42.49
2.000	0.452	-6.88	-38.45
5.000	0.298	-10.51	-29.80
10.000	0.243	-12.27	-19.33
20.000	0.224	-12.97	-10.62
50.000	0.218	-13.20	-4.35
100.000	0.217	-13.24	-2.19
200.000	0.217	-13.24	-1.01
500.000	0.217	-13.24	-0.43
1000.000	0.217	-13.24	-0.22

The quantities given in Tables 5.3 and 5.4 are plotted in Figure 5.2. These computed characteristics, based on the standard generator parameters supplied by the manufacturers, will be compared with measured parameters, determined from field tests, in a later section. The behavior of the coupled circuits are effectively displayed by the plots of Figure 5.2. These plots, usually called Bode plots, are a familiar medium to most engineers. The magnitude plots break downward at 20 dB/decade at the inverse of the denominator time constants and break upward at the inverse of the numerator time constants. For the two inductance expressions, we compute these break points shown in Table 5.5.

The breaks in the curves are quite plainly seen for the d-axis inductance, where the effect of the physical field circuit is an important influence. In the q-axis, however, the break points are closer together and the curve seems to change uniformly, without the abrupt changes in direction noted in the d-axis plot. It should be noted that the range of frequencies of interest

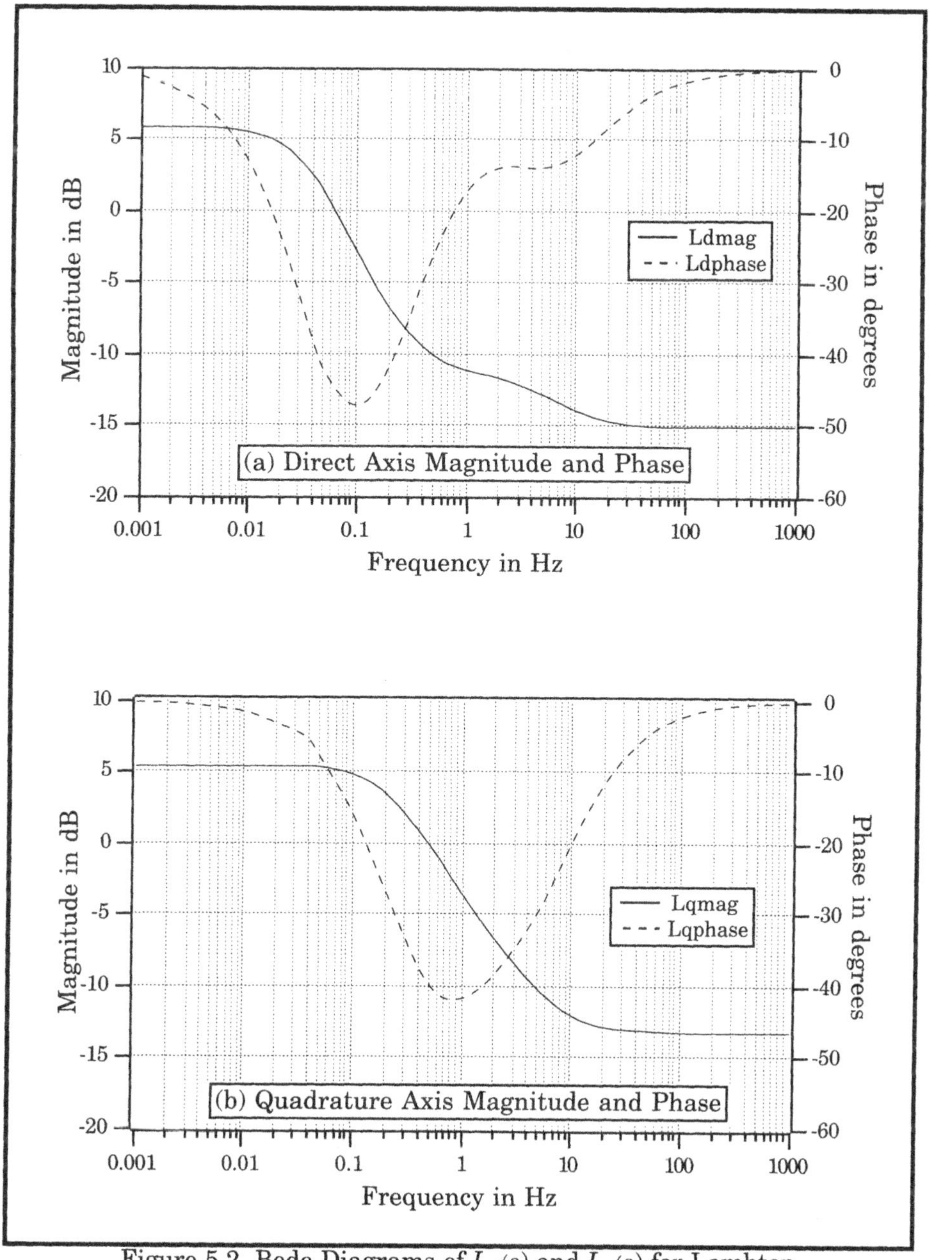

Figure 5.2 Bode Diagrams of $L_d(s)$ and $L_q(s)$ for Lambton

Table 5.5 Bode Diagram Breakpoints

Breakpoint $x = d$ or q	d axis	q axis
$1/\tau'_{xo}$	0.037	0.284
$1/\tau''_{xo}$	5.136	2.613
$1/\tau'_{x}$	0.270	1.122
$1/\tau''_{x}$	7.922	5.684

in the SSR problem is from about 10 to 50 Hz, which is near the bottom of the magnitude plot. Both the gain and the phase are changing slowly with changes in frequency in this region. This completes Example 5.1.

The foregoing example describes the kind of data that one might expect to obtain from field tests of a synchronous machine where one examines the machine characteristics under varying frequency conditions. Three methods are currently used to obtain the frequency response data from tests. These are briefly described below.

5.2.1 Standstill Frequency Response (SSFR) Tests

The SSFR test method has been extensively used by Ontario Hydro and others, and is described in [8] and [9]. Briefly, the method consists of aligning the rotor into direct or quadrature axis positions and taking measurements for impedances and transfer functions. The d and q axis test configurations are shown in Figure 5.3.

The procedure is to slowly move the rotor until a maximum coupling is observed between the stator a-phase winding and the rotor field winding. This is, by definition, the direct axis. The variable frequency voltage source is then used to excite the stator winding, with measurements made at the field winding. This is done for many frequencies, from about 0.001 Hz to 1000 Hz, for example. This gives a measure of the inductance seen looking into the stator, as well as the transfer function between stator and rotor. The process requires considerable time to complete, since each measurement at the low frequencies is very time consuming.

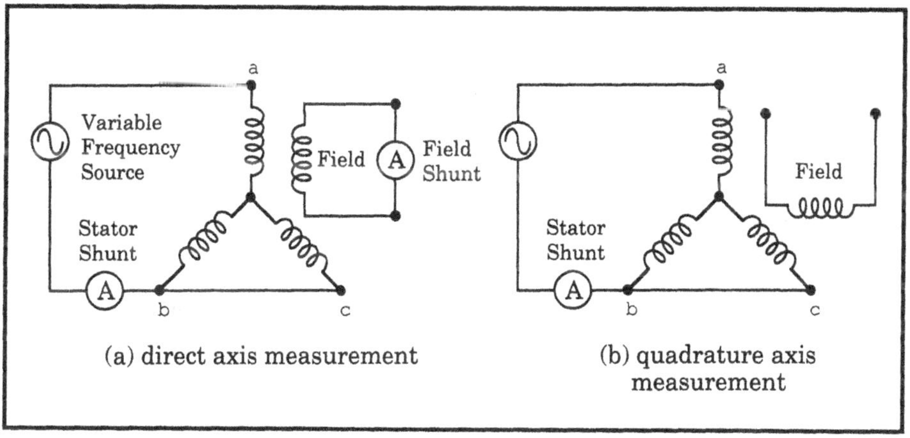

Figure 5.3 Test Configurations for Measurement of Direct and Quadrature Axis Operational Impedances and Transfer Functions

Having completed the direct axis test, the rotor is moved 90 electrical degrees and the test is repeated. For two pole machines, this requires a 90° rotation of the rotor. This configuration is, by definition, the quadrature axis position.

The SSFR tests have the following advantages:

- The tests can be performed at the factory, but this requires assembling the machine stator and rotor prior to shipment .
- The tests can be performed during the normal maintenance periods of the turbine-generator system when the generator is out of service and is readily available for testing.
- The tests are relatively inexpensive and, if performed during routine maintenance periods, do not require any generation scheduling cost or penalty.

The SSFR tests have the following disadvantages:

- The test equipment requires the use of a very linear, very high power operational amplifier to provide the variable frequency input voltage.
- Tests are usually conducted at unsaturated conditions because of inadequate test equipment power rating.

- The testing time for low frequency performance is rather long, with each point on the frequency curve requiring minutes to hours of testing time.

- Testing must be scheduled during planned maintenance outages of the generator when the machine is available for standstill observation.

- The centrifugal force effects on slot wedges are not accounted for in SSFR testing, and the error introduced by this omission is difficult to assess.

In practice, the requirement of a very high power voltage source presents a difficult problem. One reported test configuration used a 12 kW, water cooled operational amplifier that weighed over 2000 pounds [10]. Even this amplifier was incapable of producing saturation in a large generator, and there was considerable doubt that it produced currents that were even in the linear portion of the saturation curve, but were, instead, on the toe of the curve. Despite the disadvantages of the SSFR method, however, it is one of the few methods for obtaining q-axis data directly, and the method has a number of proponents for this reason. Recent tests have confirmed that good machine data may be obtained by this method for most machines. Some machines, particularly those with pole face amortisseur windings, have not been modeled as accurately using this method as machines without such damper circuits.

The standstill frequency response testing procedures have matured considerably in recent years and are now described in IEEE Standard 115A [4]. Additional references on the standstill frequency response testing method and results are given in [11-15].

5.2.2 Generator Tests Performed Under Load

Another type of generator testing that has been used successfully is quite different from the stand still tests in that the generator is connected to the system, and operating in a more or less normal condition. Three of these "under load" tests will be described.

5.2.2.1 The On-Line Frequency Response Test

The On-Line Frequency Response (OLFR) test was first reported by engineers at Ontario Hydro in 1980 [16]. The OLFR tests are conducted with the machine operating at rated speed; usually with the machine on line, and carrying about 80% load. The machine frequency response is obtained by applying sinusoidal signals to the voltage regulator summing junction and measuring the steady state changes in the field voltage and current, the rotor speed, the terminal voltage, and both the active and reactive power output. Any power system stabilizer, if present, is removed, and the voltage regulator gain is set to a low value prior to injecting the sinusoidal signals.

The analytical procedure for determining the desired transfer functions is based on the development of a state-space description of the machine using well known techniques. An identification algorithm is then applied to minimize the error between the measured and the computed responses at a number of points in the specified frequency range. The technique is an iterative one, which is continued until the sum of the squares of the errors becomes a minimum.

These are relatively expensive tests, since they may require special scheduling of the unit, and are difficult to perform. However, these tests have the advantage that saturation effects are included in the test data. They also provide a method of including the rotational effect of such components as slot wedges on the generator inductances and time constants. Additional information on the OLFR test method may be found in [11].

5.2.2.2 Load Rejection Test

Another test that is conducted with the generator operating is the Load Rejection test, which was first proposed by deMello in 1977 [17]. The procedure is to operate the machine connected to the power system with as near to zero power as possible and with the excitation system on manual control. In this predisturbance condition, the generator can be established at any desired excitation, either over or underexcited. The test disturbance is to open the generator circuit breaker and to record the transient decay in terminal voltage and field current.
.i.synchronous machine parameters:load ;rejection tests;
When underexcited conditions are established, the machine will not be in a highly saturated condition, thereby establishing the condition for finding the unsaturated values of direct axis reactances and time constants, which are found from the decaying plots of terminal voltage and field current following the load rejection. Repeating the test with the generator over-excited will then provide saturated values for these direct axis parameters.

To determine the quadrature axis parameters, a novel concept is introduced whereby the armature current phasor lines up exactly with the quadrature axis. Then the value of x_q is determined by knowing only the reactive power and the armature current [17]. This desired loading condition is found by performing successive load rejection runs where the field current is monitored, with the objective of reaching the condition where there is no transient deviation in field current. With the proper condition established, it can be shown that the quadrature axis reactances can be computed.

The load rejection method provides another method for determining the quadrature axis inductances. The method is somewhat difficult to implement, and requires rescheduling of the unit, which is costly in most cases. It also requires several load rejection trips of the unit, which may not be well received by the plant management.

5.2.2.3 Off-Line Frequency Domain Analysis of Disturbances

This method differs from the On-Line Frequency Response tests in that the frequency domain analysis is performed off-line and the test disturbance is a line switching event, rather than the injection of signals into the voltage regulator. The parameters obtained therefore include the effects of saturation, and the machine operating condition is a normal loaded one.

The technique requires the recording of various generator terminal conditions, such as voltage, current, and speed, during the switching event. These recorded data are processed off-line to develop digital representations of the raw data, from which v_d, v_q, i_d, and i_q are derived and their Laplace transforms computed. From these transforms, the usual Bode plot of $L_d(s)$ and $L_q(s)$ are formed. The test results from this technique appear to be quite acceptable, based on computer simulations of the disturbance events.

5.2.3 Other Test Methods

Several other tests have been devised for obtaining generator data. These are described below.

5.2.3.1 The Short Circuit Test

Another variation of testing with the machine operating is the short circuit test [21, 22]. In this test, the machine is running at reduced speeds, with a line-to-line short circuit between phases, and with excitation applied briefly to produce line-to-line short circuit currents at fundamental frequency corresponding to the running speed. The records of the short circuit currents, together with records of the rotor angle, are then processed off line to yield the d and q components of voltages, currents, and flux linkages. Fourier analysis is performed on these results to provide the operational inductances as a function of frequency.

The unique feature of the method is the use of the line-to-line short circuit current, which has a simple relationship to the *d* and *q* axis components of current [23]. The method would be difficult to implement if the excitation system is a rotating system, but would be feasible if static excitation is used. There would be practical difficulties in arranging for the fault connection, and the plant management may object to such a procedure.

One of the advantages claimed for this method is that it provides the machine characteristics for stability, in the 0 to 10 Hz range, but it also gives information that is useful in determining machine characteristics in the range of frequencies for SSR.

5.2.3.2 Trajectory Sensitivity Based Identification

This method utilizes the measured data taken from a machine as it responds to system events such as faults or line trippings [24]. The identification technique is based on the use of trajectory sensitivity and least squares to compute changes in the model parameters to minimize an error function.

The trajectory sensitivity method involves the following steps in its implementation:

1. Postulate the model structure and determine the parameters to be identified.
2. Make an initial guess for the unknown parameters.
3. Use the measured data to compute the output of the model.
4. Stop if the simulated output is sufficiently close to the measured output.
5. Use the measured input data to simulate the output of the trajectory with respect to the unknown parameters.
6. Compute an adjustment to the unknown parameters.
7. Update the unknown parameters.
8. Go to #3 and repeat 3 through 8.

The authors of the method report very good results. It was also noted that some of the parameters were quite different from the initially assumed values.

5.3 PARAMETER FITTING FROM TEST RESULTS

The steps involved in calculating model parameters from the test data are briefly described below. The parameter fitting is a recursive process that is well suited to digital computer evaluation. Briefly, the procedure is as follows:

1. Choose the number of rotor circuits in the d- and q-axes of the machine model.

2. Approximate the measured inductances and transfer functions as the ratio of polynomials of the same order as the number of rotor circuits selected in Step 1.

3. Start with some approximate value of the parameters, such as those derived using equations (5.1) to (5.5) that are taken from conventional stability data.

4. Calculate the coefficients of the polynomials using the approximate model and compare with coefficients of the polynomial of Step 2.

5. Adjust the overall model parameters using some algorithm like a Newton-Raphson technique.

6. Repeat Steps 4 and 5 until the error is acceptably small.

7. As a final check, calculate the frequency response from the derived model parameters and compare with the test result.

The algorithm can be modified to obtain a closer fit to the measured data in the frequency range of interest. This is done by matching the measured frequency response with the calculated frequency response, rather than matching the coefficients of the polynomial derived from a least square fit to the measured frequency response. Another common approach is to use the SSFR data to obtain an approximate model and then use the OLFR data to further improve the model. Reference [16] uses this technique.

5.4 SAMPLE TEST RESULTS

A few sample test results from [11] are reproduced here with a brief explanation of each. Results for two machines, one without a pole face amortisseur, and one with a pole face amortisseur, are presented. The machines are both owned by Ontario Hydro and are identified as follows[1]:

[1]Permission by Ontario Hydro to reproduce these results is gratefully acknowledged.

- The Lambton Generator:
  500 MVA, 0.85 pf, 3600 rpm
  No Pole Face Amortisseur

- The Nanticoke Generator:
  500 MVA, 0.85 pf, 3600 rpm
  Full Length Pole Face Amortisseur.

Figures 5.4(a) and 5.4(b) show the SSFR measured magnitude and phase of the quantities indicated for the Lambton generator and compare these measured quantities with both two and three rotor circuit model curve fits. Figures 5.5(a) and 5.5(b) show similar results for the Nanticoke unit. These two machines have been studied extensively and a great deal of information is available about them (See [8-13], for example).

In the above figures the following notations are used.

SSFR2 Standstill frequency response data fitted to the two rotor circuit model;

SSFR3 Standstill frequency response data fitted to the three rotor circuit model;

OLFR2 On-line frequency response data fitted to the two rotor circuit model;

OLFR3 On-line frequency response data fitted to the three rotor circuit model.

Figure 5.6 shows the three rotor circuit model used in this experiment. The two rotor circuit model is a simplification of this model with the third rotor circuit omitted. Note that the symbols are different than those used in this report. These symbols are not standardized and different symbols appear in the literature. More important than the symbols is the addition of two new so-called leakage inductances subscripted $fkd1$ and $fkd2$. These leakages have been found necessary and are testimony that the constant mutual flux linkage model is inadequate. Moreover, these new leakages have been shown to change with saturation, and even result in negative inductance values in some cases.

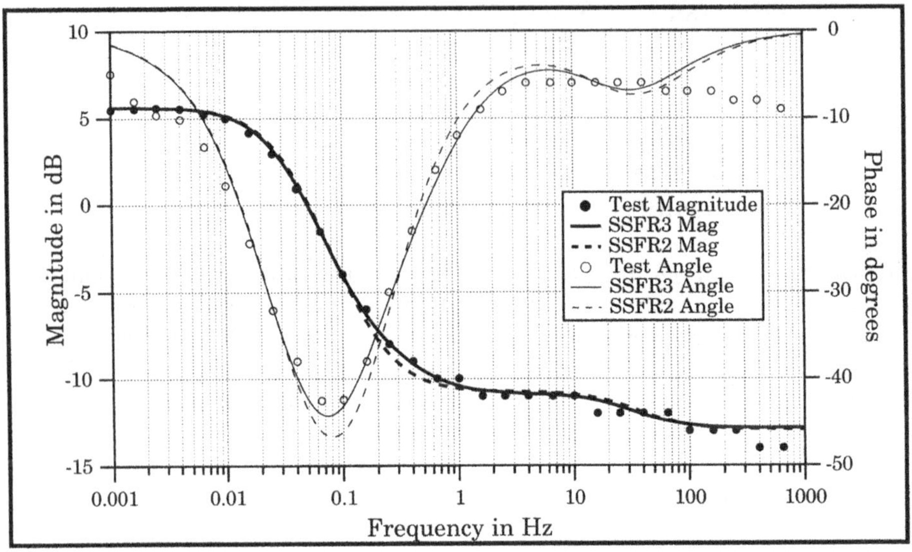

Figure 5.4 (a) Two and Three Rotor Winding Fits of Lambton d-axis Operational Inductance (Data from [10] with permission)

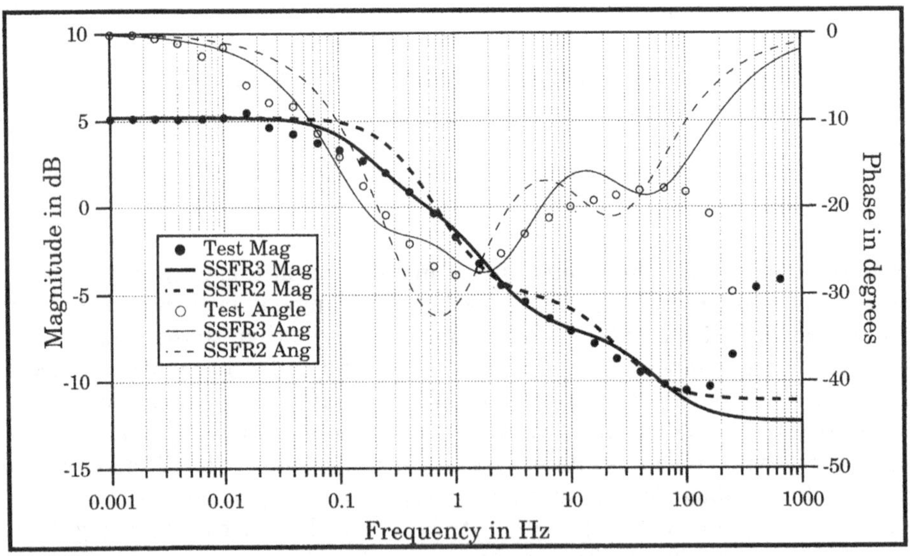

Figure 5.4 (b) Two and Three Rotor Winding Fits of Lambton q-axis Operational Inductance (Data from [10] with permission)

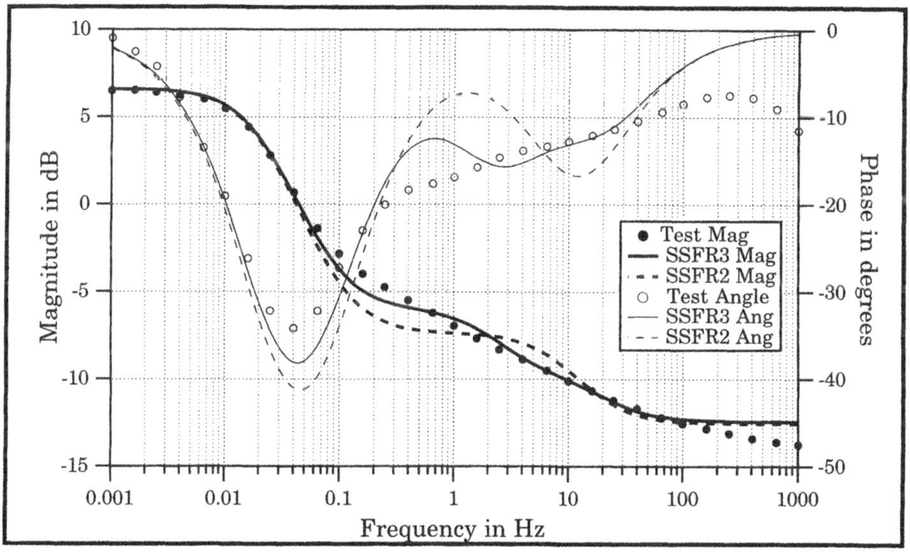

Figure 5.5 (a) Two and Three Rotor Winding Fits of Nanticoke d-axis Operational Inductance (Data from [10] with permission)

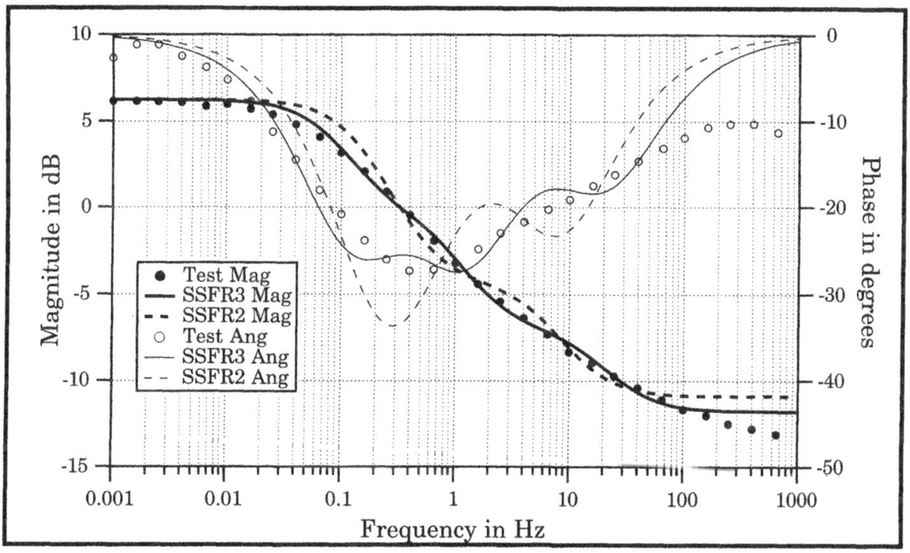

Figure 5.5 (b) Two and Three Rotor Winding Fits of Nanticoke q-axis Operational Inductance (Data from [10] with permission)

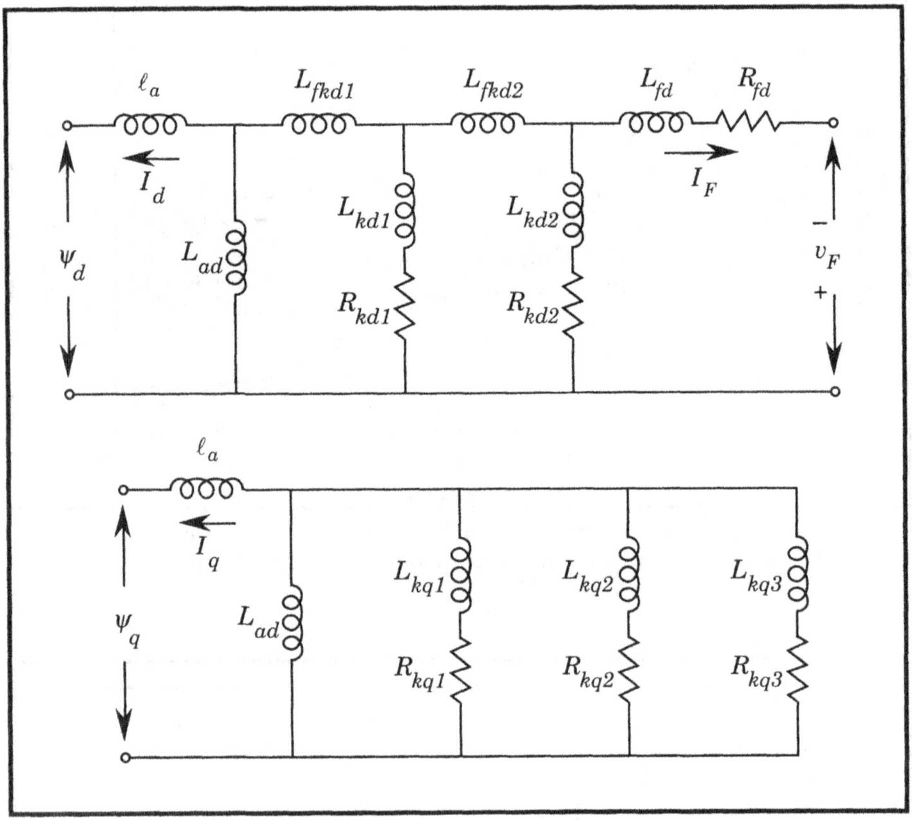

Figure 5.6 Machine Equivalent Circuit used in the Curve Fitting Procedure of Ontario Hydro

Tables 5.6 and 5.7 compare the model parameters for the various models for the two machines, Lambton and Nanticoke of Ontario Hydro. The column labeled "Standard Model" shows the data normally supplied by the manufacturer.

The proof of any model is its comparison with a field test. Such a model validation was conducted by Ontario Hydro for the two units under study by performing line switching tests under carefully observed test conditions.

The results of these field validation tests are compared with the computed response of the various models as shown in Figures 5.7 for Lambton and 5.8 for Nanticoke. Observe that the Lambton generator is modeled reasonably

Table 5.6 Standard and Derived Equivalent Circuit Parameters for the Lambton Synchronous Machine

Model Parameter	Standard Model	SSFR2 Parameter	OLFR2 Parameter
L_{AD}	1.810	1.858	1.858
L_{AQ}	1.707	1.762	1.845
ℓ_a	0.160	0.155	0.142
L_{fd}	0.1171	0.01051	0.01119
R_{fd}	0.001189	0.001084	0.001084
L_{kd1}	0.01737	0.01136	0.01102
R_{kd1}	0.0109	0.01065	0.0100
L_{fkd1}	-----	0.1328	0.1562
L_{kq1}	0.0638	0.5677	0.4502
R_{kq1}	0.0164	0.0147	0.0100
L_{kq2}	0.3833	0.1717	0.1983
R_{kq2}	0.0099	0.1765	0.1989

well using the SSFR2 model and the improvement over the standard data model is evident. The SSFR2 model has proper damping but the frequency is still in error. The Nanticoke model, however, is not modeled accurately for damping using even the SSFR3 model. Here, the additional model tuning using the OLFR test data is shown to be necessary.

Unfortunately, no tests have been conducted to validate any model for use in SSR studies. Even so, there is no reason why the techniques used to improve the model for stability studies would not give good results for SSR models. The results of the curve fitting technique suggest that separate models for SSR are desirable and may eventually be found to be necessary.

The second and third order models serve fairly well for transient stability, where the frequencies of interest are generally less than five Hz. An examination of Figures 5.4 and 5.5 in the 10 to 50 Hz region indicates that even a third order model is a poor fit, particularly on the q axis. The real question becomes one of determining exactly how accurate the model must be to obtain study results that are adequate and conservative. This question

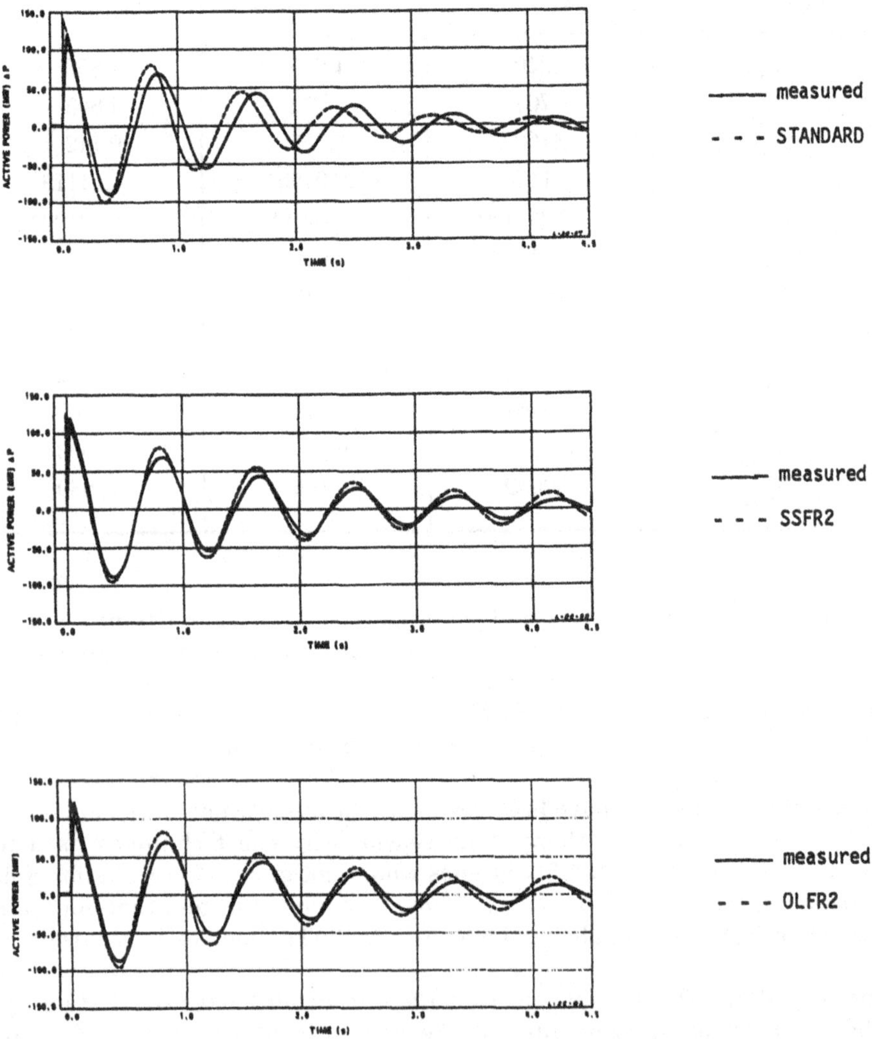

Figure 5.7 Lambton Line Switching Test Active Power Record Compared with Simulated Results Using Standard, SSFR2, and OLFR2 Machine Models [11]

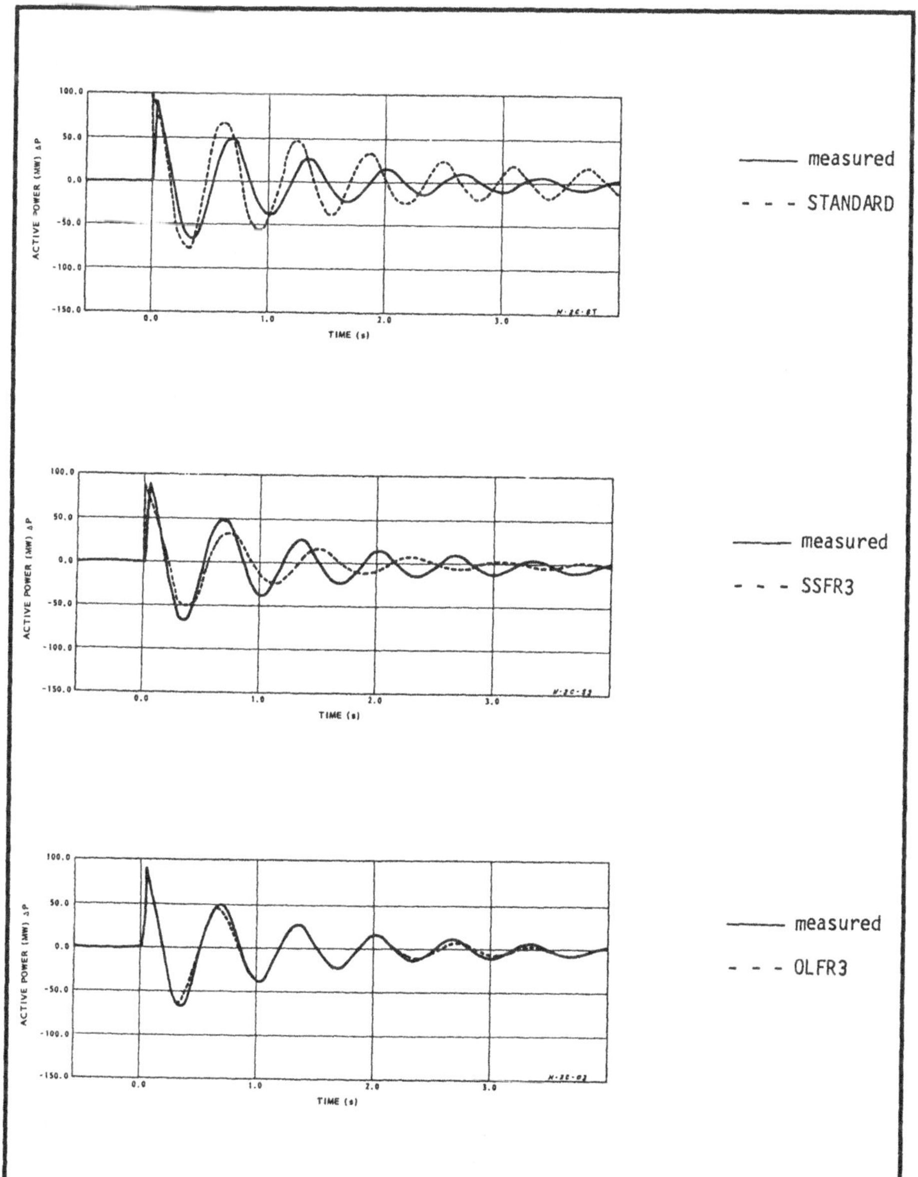

Figure 5.8 Nanticoke Line Switching Test Active Power Record Compared with Simulated Results Using Standard, SSFR3, and OLFR3 Machine Models [11]

Table 5.7 Standard and Derived Equivalent Circuit Parameters for the Nanticoke Synchronous Machine

Model Parameter	Standard Model	SSFR2 Parameter	OLFR2 Parameter
L_{AD}	2.165	2.152	2.152
L_{AQ}	2.047	2.057	2.057
ℓ_a	0.195	0.172	0.172
L_{fd}	0.0908	0.0155	0.2785
R_{fd}	0.00076	0.00094	0.00083
L_{kd1}	0.0457	2.732	5.182
R_{kd1}	0.00703	0.1142	0.0969
L_{fkd1}	-----	-0.5215	-0.0403
L_{kq1}	0.1560	1.657	1.4475
R_{kq1}	0.00390	0.00538	0.00433
L_{kq2}	0.0378	0.1193	0.0560
R_{kq2}	0.00139	0.1081	0.0122
L_{fkd2}	-----	0.8975	-----
L_{kd2}	-----	0.00753	0.0369
R_{kd2}	-----	0.00592	0.0130
L_{kq3}	-----	0.4513	0.4064
R_{kq3}	-----	0.0188	0.0017

has not been answered by past research and remains a valid concern for future studies. This suggests that it is prudent to use a study margin in devising SSR countermeasures to account for possible inaccuracies in the modeling.

5.5 FREQUENCY DEPENDENT *R* AND *X* DATA

Occasionally, data is available from the manufacturer in the form of R and X as a function of frequency. These are usually calculated values that the manufacturer computes using a detailed higher order model based upon design information and past experience. The available data can then be mathematically represented by equivalent circuits, as shown in Figure 5.9.

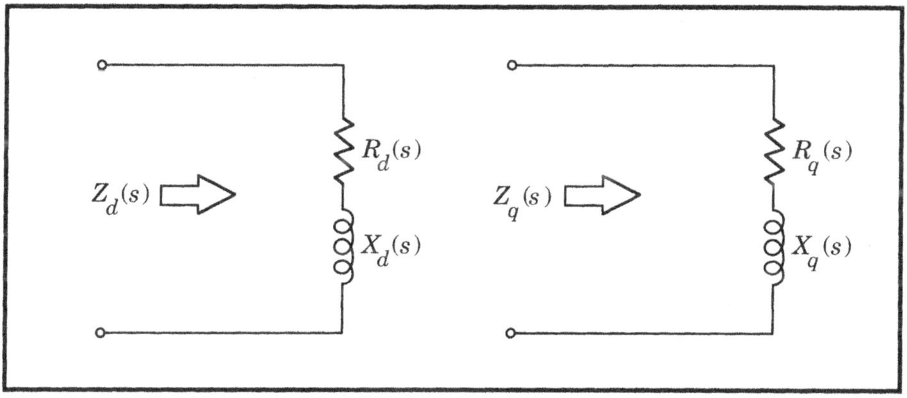

Figure 5.9 *D* and *Q* Axis Frequency Dependent Equivalents

The data is supplied either in a tabular form or in the form of curves. To use these data in an eigenvalue computation one needs to find model parameters. As discussed in previous section regarding the treatment of measured data, the following steps are involved:

- Select the order and type of model;
- Curve fit the supplied data with the corresponding order polynomials;
- Find the model parameters by comparing the coefficients of polynomials using a recursive method until the error is acceptably small;
- Adjust the model parameter in the frequency range of interest by comparing the frequency response calculated from the model to that supplied in the *R* and *X* vs Frequency data.

Unfortunately, the model parameters obtained using the above approach will not be unique. In fact, an infinite number of combinations of model parameters exist that can produce approximately the same response. A further complication is introduced if, instead of $L_d(s)$ and $L_q(s)$ being available separately, only the average value is available. Since there is no fixed ratio between the model parameters of the d-axis and q-axis, it is not practical to attempt to derive the model parameters from the average value of *R* and *X* vs Frequency.

5.6 OTHER SOURCES OF DATA

In some cases, the complete model parameters may be available directly from the manufacturer or some other source. Knowing the model, the model parameters can be used to compute the transfer functions $L_d(s)$, $G_d(s)$, and $L_q(s)$ for use in the eigenvalue program. For the two rotor-circuit equivalent model of Figure 5.1, the relationship among the model parameters and coefficients of the transfer functions are given in Table 2.3. Similar relationships can be derived for any other model.

Synchronous machine modeling and the determination of machine model parameters continues to be a problem of considerable interest, with new results being published regularly. The references provide additional information on this subject.

5.7 SUMMARY

In this chapter, we have presented some of the special needs for machine data in the study of SSR. Data normally used for transient stability is usually considered quite adequate for the 0 to 5 Hz bandwidth that is important for transient stability, but this same data may not be adequate for the study of SSR. This is graphically illustrated in Figures 5.4 and 5.5, where the rather large phase errors in the SSR frequency range are apparent. In many cases, however, the transient stability data is all that is available. In defense of using these data, it must be acknowledged that no turbine-generator shafts have been known to have failed based on the use of stability data for the design of SSR countermeasures.

The authors recommend that the engineer discuss the SSR data requirements with the machine manufacturer. In many cases special data can be provided for SSR studies, once the study requirements are known.

Another important source of information is a new IEEE "Guide for Synchronous Generator Modeling Practices in Stability Analysis," which is in preparation as this book is being written and should be balloted upon during late 1989 or early 1990. Although written specifically for transient stability modeling practices, much of this document is valuable for SSR as well.

5.8 REFERENCES FOR CHAPTER 5

1. IEEE Joint Working Group on Determination of Synchronous Machine Stability Study Constants, "Synchronous Machine Stability Constants--Requirements and Realizations," IEEE-PES paper A 77-210-8, presented at the PES Winter Meeting, New York, 1977, abstract in **IEEE Trans.**, v. PAS-96, n. 4, July-Aug 1977, p. 1076.

2. IEEE Committee Report, "Supplementary Definitions and Associated Test Methods for Obtaining Parameters for Synchronous Machine Stability Study Simulations," ibid, v. PAS-99, July/August 1980.

3. IEEE Symposium Record, "Symposium on Synchronous Machine Modeling for Power System Studies," Presented at the IEEE PES Winter Meeting, Feb. 2, 1983, New York, IEEE Pub. No. 83THO101-6 PWR.

4. IEEE Std. 115A-1983, "IEEE Trial Use Standard Procedures for Obtaining Synchronous Machine Parameters by Standstill Frequency Response Testing," IEEE, New York, 1983.

5. IEEE P1110/D8, "Guide for Synchronous Generator Modeling Practices in Stability Analysis," Draft standard in preparation, January 1989.

6. IEEE Subsynchronous Resonance Working Group, "Second Benchmark Model for Computer Simulation of Subsynchronous Resonance," ibid, v. PAS-104, n. 5, 1985, p. 1057-1066.

7. Adkins, Bernard, **The General Theory of Electrical Machines**, Chapman & Hall, 1962.

8. Coultes, M. E., and W. Watson, "Synchronous Machine Models by Standstill Frequency Response Tests," ibid, v. PAS-100, April 1981, p. 1480.

9. Dandeno, P. L., and A. T. Poray, "Development of Detailed Turbogenerator Equivalent Circuits from Standstill Frequency Response Measurements," ibid, v. PAS-100, April 1981, p. 1646.

10. NEI Parson Report, "Determination of Synchronous Machine Stability Study Constants," EPRI Report EL-1424, v. 4, EPRI Project 997-1, August 1980.

11. Ontario Hydro Report, "Determination of Synchronous Machine Stability Study Constants," EPRI Report EL-1424, v. 2, EPRI Project 997-2, December 1980.

12. Watson, W. and G. Manchur, "Synchronous Machine Operational Impedances from Low Voltage Measurements at the Stator Terminals," **IEEE Trans.**, v. PAS-93, May/June 1974, p. 777.

13. EPRI Research Project 997, "Determination of Synchronous Machine Stability Study Parameters," v.1 (Westinghouse), v.2 (Ontario Hydro), v.3 (Power Technologies, Inc.), v.4 (NEI Parsons), EPRI Report EL-1424.

14. deMello, F. P., L. N. Hannett, and J. R. Willis, "Determination of Synchronous Machine Stator and Field Leakage Inductances from Standstill Frequency Response Tests," IEEE paper 88 WM 158-8, presented at the IEEE Power Engineering Society Winter Meeting, January 31-February 5, 1988, New York.

15. Boje, E. S., J. C. Balda, R. G. Harley, and R. C. Beck, "Time-Domain Identification of Synchronous Machine Parameters from Simple Standstill Tests," IEEE paper 88 WM 003-6, presented at the IEEE Power Engineering Society Winter Meeting, January 31-February 5, 1988, New York.

16. Dandeno, P. L., P. Kundur, A. T. Poray, and H. M. Zein-el-Din, "Adaption and Validation of Turbogenerator Model Parameters Through On-Line Frequency Response Measurements," **IEEE Trans.**, v. PAS-100, April 1981, p. 1656.

17. deMello, F. P., and J. R. Ribeiro, "Derivation of Synchronous Machine Parameters from Tests," **IEEE Trans.**, v. PAS-96, July/August 1977, p. 1211-1218.

18. deMello, F. P. and L. H. Hannett, "Validation of Synchronous Machine Models and Derivation of Model Parameters From Tests," **IEEE Trans.**, v. PAS-100, n. 2, 1981, p. 662-672.

19. Sugiyama, T., T. Nishiwaki, S. Takeda, and S. Abe, "Measurements of Synchronous Machine Parameters Under Operating Condition," IEEE paper 81 SM 428-2, presented at the IEEE Power Engineering Society Summer Meeting, Portland, 1981.

20. deMello, F. P., L. N. Hannett, D. Smith, and L. Wetzel, "Derivation of Synchronous Machine Stability Parameters from Pole Slipping Conditions," **IEEE Trans.**, v. PAS-101, n. 9, 1982, p. 3394-3402.

21. deMello, F. P. and L. N. Hannett, "Determination of Synchronous Machine Electrical Characteristics by Test," **IEEE Trans.**, v. PAS-102, n. 12, 1983, p. 3810-3815.

22. Eitelberg. E., and R. G. Harley, "Estimating Synchronous Machine Electrical Parameters from Frequency Response Tests," IEEE paper 86 WM 208-3, presented at the IEEE Power Engineering Society Winter Meeting, New York, 1986.

23. Concordia, C., **Synchronous Machines**, John Wiley and Sons, New York, 1951.

24. Sanchez-Gasca, J. J., C. J. Bridenbaugh, C. E. J. Bowler, and J. S. Edmonds, "Trajectory Sensitivity Based Identification of Synchronous Generator and Excitation System Parameters," IEEE paper 88 WM 205-7, presented at the IEEE Power Engineering Society Winter Meeting, New York, January 1988.

25. IEEE Subsynchronous Resonance Working Group, "First Benchmark Model for Computer Simulation of Subsynchronous Resonance," ibid, v. PAS-99, September/October 1980.

26. Sharma, D. K., D. H. Baker, J. W. Dougherty, M. D. Kankam, S. H. Minnich, and R. P. Schulz, "Generator Simulation Model Constants by Finite Elements: Comparison With Test Results," ibid, v. PAS-104, July 1985, p. 1812-1821.

27. Crappe, M., M. Delhaye, M. Naciri, Ph. Lorent, and L. Soenen, "Experimental Determination of Large Turbogenerator Dynamic Parameters by Computer Aided Analysis," CIGRE Paper 84-Presented at the International Conference on Large High Voltage Electric Systems, August-September 1984, Paris.

CHAPTER 6

TURBINE-GENERATOR SHAFT MODEL PARAMETERS

The preceding chapter discusses the problems inherent in determining accurate parameters for the synchronous machine. This chapter presents similar information regarding data for the turbine-generator shaft model. As with the synchronous machine, there are two sources of data. The first is the data provided by the manufacturer, and this is always the best place to start. But, as with the synchronous machine data, the manufacturer's shaft data may require validation by field testing to assure accurate representation in SSR simulation studies.

There are two ways of representing turbine-generator shaft parameters in the eigenvalue program, namely

- A shaft spring-mass model;
- A modal model.

Both representations are described in this chapter.

6.1 THE SHAFT SPRING-MASS MODEL

Generally, the spring-mass model of the turbine-generator shaft is known with relatively good accuracy. The manufacturers supply the spring-mass data based on design values and have, in the past, been able to predict the torsional frequencies within about one Hz for the significant modes. If the torsional frequencies are known more accurately from tests, then the data can be revised to match the measured frequencies. The manufac- turer-supplied data can be easily converted into the units that are needed for analytical studies such as eigenvalue analysis.

Consider the spring-mass shaft model shown in Figure 6.1, where we consider the resilient shaft sections connecting the masses to be linear springs, and with damping between adjacent masses and from each mass to the stationary reference. We usually assume that these damping parameters are viscous, i.e., the retarding torque produced by the damping is directly proportional to the speed of relative motion between the parts. The least accurately known components of the spring-mass parameters are the damping elements. Estimated values, based on current estimation

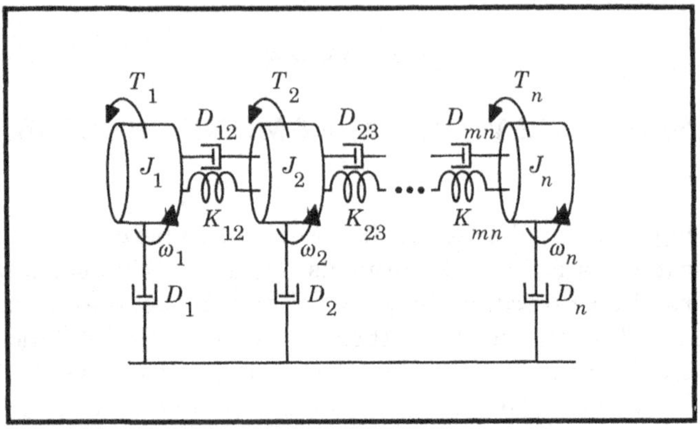

Figure 6.1 Spring-Mass Model for an N Mass Shaft

techniques, have been found to differ from measured values by as much as a factor of two to five. Even though this difference is becoming smaller with experience, the state-of-the-art is not advanced enough to predict damping with great confidence.

If the spring-mass model is used in an eigenvalue or other analytical program, the damping needs to be provided in the form of .i.dashpot; (viscous) damping for each element and the mutual damping between adjacent elements, as shown in Figure 6.1. The measured or predicted dampings provided by the manufacturers, on the other hand, are often in modal form and are not suitable as direct input data to the spring-mass model. There are two ways of handling this problem – neglect damping or approximate the damping effect. Both methods are discussed below.

6.1.1 Neglecting The Shaft Damping

The turbine-generator mechanical damping coefficients are generally too small to have any impact on the torsional frequencies. This is equivalent to asserting that the damped and undamped natural frequencies are equal, or are very close to being equal (within 0.1 Hz).

This can be illustrated by a simple example. Consider a single flywheel, like a large doorknob, connected to a wall (the door) by a shaft, as shown in Figure 6.2. Let the flywheel have moment of inertia J, and the shaft have spring constant K and damping D. Now mark the flywheel when at rest with no external forces acting on it so that we can measure θ, the angle of deflection, and visualize that it as positive in a given direction, such as the counter-clockwise direction. The Newton equation of motion of this system is given by

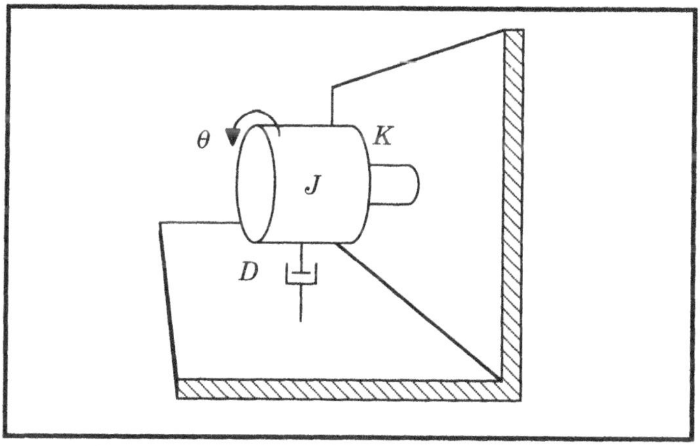

Figure 6.2 A Single Mass Rotating System

$$J\frac{d^2\theta}{dt^2} = T - D\omega - K\theta \tag{6.1}$$

where T is an externally applied torque trying to twist the doorknob in the positive direction. Since our flywheel-doorknob is fixed, it is not able to turn, but the shaft can twist due to the resilience of the material. If we solve (6.1) in the Laplace domain, we get

$$\theta(s) = \frac{T(s) + (Js + D)\theta(s) + J\dot{\theta}(0)}{Js^2 + Ds + K} \tag{6.2}$$

where

$$\theta(0) = \text{Initial Value of the Angle}$$
$$\dot{\theta}(0) = \omega(0) = \text{Initial Angular Velocity}$$

Now let

$$T = 0 \tag{6.3}$$
$$\omega(0) = 0 \tag{6.4}$$
$$\theta(0) = \text{A Constant.} \tag{6.5}$$

Condition (6.3) asserts that there is no external torque applied. Condition (6.4) requires that the system is initially at rest. Finally, (6.5) indicates that we initially hold the doorknob such that the shaft is slightly twisted and let go at time $t = 0$. We are interested in the ensuing oscillations, which depict the properties of the spring-mass model. The solution is given by

$$\theta(t) = Ae^{-\alpha t}\cos\omega_r t. \tag{6.6}$$

We analyze this result by first rewriting the system equation as follows, including the given boundary conditions:

$$\theta(s) = \frac{(s + D/J)\theta(0)}{s^2 + (D/J)s + (K/J)} \tag{6.7}$$

or we rewrite (6.7) as

$$\theta(s) = \frac{(s + 2\zeta\omega_n)\theta(0)}{s^2 + 2\zeta\omega_n s + \omega_n^2} \tag{6.8}$$

where we define

$$\begin{aligned} \zeta &= \text{Damping Factor} \\ \omega_n &= \text{Undamped Natural Frequency.} \end{aligned} \tag{6.9}$$

From this system, from (6.7) and (6.8), we compute

$$\zeta = \frac{1}{2}\sqrt{\frac{D^2}{JK}} \tag{6.10}$$

$$\omega_n = \sqrt{\frac{K}{J}} \tag{6.11}$$

$$\alpha = \zeta\omega_n = \frac{D}{2J} = \text{System Damping} \tag{6.12}$$

$$\omega_r = \omega_n\sqrt{1 - \zeta^2} = \text{Damped Frequency} \tag{6.13}$$

or

$$\omega_r = \sqrt{\frac{K}{J} - \frac{D^2}{4J^2}}. \tag{6.14}$$

For most physical shaft systems D is very small, hence ζ is also small. This means that the damped and undamped frequencies are nearly equal. Note that for $D<1$ and $J>1$, (6.11) is very nearly equal to (6.14).

If the damped and undamped natural frequencies are nearly equal, then the eigenvalue calculation can be made with no damping represented with little error. The real part of the computed eigenvalue will then represent the negative damping due to torsional interaction. The effect of mechanical dampings (predicted or measured) can then be accounted for by algebraically adding these modal dampings to the calculated negative damping. The eigenvalue or other analysis can, therefore, be designed to take the modal dampings as input parameters and account for them in the calculated eigenvalues at the end of the computation. The effect of this accounting will be transparent to the user.

The terminology that is often used in the discussion of system damping should be noted. The damping term in the equations is the parameter that multiplies the first derivative term (or the s term in the Laplace domain). It is sometimes called the "damping factor" or "damping coefficient" in the system characteristic equation. In most physical systems, at least under normal conditions, this coefficient is a positive coefficient and, if this is true, the response of the second order system will be "damped," that is, the oscillations will gradually die out. It is sometimes said that the system is "positively damped" or has "positive damping," although the *positive* qualification could be considered redundant since the term *damp* means to decrease in amplitude.

In SSR, and some other physical systems, the damping coefficient may be negative, which results in growing oscillations. This is sometimes referred to in the literature as "negative damping" or "undamping," with the two terms both meaning that the response is *not damped*. It is unfortunate that there is no word that expresses this condition succinctly so the contradictory term "negative damping" could be eliminated. In this book, we reluctantly follow the common practice and use both "negative damping" and "undamping," since there seems to be no alternative.

6.1.2 Approximate Damping Calculations

Theoretically, any set of modal dampings can be converted to equivalent dashpot self and mutual dampings by means of reverse transformation from the modal domain to the spring-mass domain. Unfortunately, such a reverse transformation leads to mutual damping between not only the adjacent elements, but among all elements irrespective of their physical location on the shaft. The spring-mass model, generally, does not provide for these fictitious mutual damping elements, as noted in Chapter 4, but represents mutual damping only between adjacent elements. Either the spring-mass model will have to be modified to allow these fictitious damping elements, or a parameter fitting technique must be developed to compute the elemental damping coefficients from the modal dampings.

A second alternative would be to accurately represent only one of the modal dampings. Often, in an SSR analysis, one knows the critical SSR mode that is of greatest concern. If this is the case, a set of dashpot dampings that are proportional to the inertias can be found such that they accurately represent the damping of one mode. No mutual damping is required. Thus the modal damping for any one mode can be represented accurately, but the modal damping of all other modes will be in error. Generally, these errors are small enough so as not to cause any significant concern in the resulting calculation.

We pursue here a third alternative for adjusting the spring-mass model to agree as closely as possible with the measured modal dampings, which we shall call "model adjustment."

6.1.2.1 Model Adjustment

The spring-mass model can be adjusted to simulate the measured modal dampings and frequencies as long as the damping values and frequency adjustments are relatively small. An approximate method is described for making these adjustments.

The dynamics of the turbine-generator shaft shown in Figure 6.1 are given by the following vector matrix equation.

$$\mathbf{J}\ddot{\theta} + \mathbf{D}\dot{\theta} + \mathbf{K}\theta = \mathbf{T} \tag{6.15}$$

where

$\mathbf{J}$ = an n × n diagonal inertia matrix
$\mathbf{D}$ = an n × n diagonal damping matrix
$\mathbf{K}$ = an n × n nondiagonal spring constant matrix.

For this analysis, we are neglecting the damping between adjacent masses and concentrating only on the damping from each mass to the reference. The damping between adjacent masses is smaller than the damping to reference and can often be neglected.

If **K** were also a diagonal matrix, (6.15) would represent n decoupled second order differential equations that could be solved independently. Fortunately, there exists a linear transformation that completely decouples the system (6.15). The transformation matrix is the matrix consisting of the mode shapes (eigenvectors) as columns (or rows), and this matrix is constructed by a procedure exactly analagous to that used for the synchronous machine, which resulted in the transformation **Q** given by (2.33). We shall again let **Q** be the transformation matrix and **δ** be the new vector of coordinate system variables. Then the following equations define the relationship between the coordinate systems.

$$\boldsymbol{\theta} = \mathbf{Q}\boldsymbol{\delta} \tag{6.16}$$

$$\dot{\boldsymbol{\theta}} = \mathbf{Q}\dot{\boldsymbol{\delta}} \tag{6.17}$$

$$\ddot{\boldsymbol{\theta}} = \mathbf{Q}\ddot{\boldsymbol{\delta}} \tag{6.18}$$

We emphasize that both **θ**and **δ** are $n \times 1$ vectors of shaft angles.

Substituting for $\boldsymbol{\theta}$, $\dot{\boldsymbol{\theta}}$, and $\ddot{\boldsymbol{\theta}}$ we compute

$$\left(\mathbf{Q}^T\mathbf{J}\mathbf{Q}\right)\ddot{\boldsymbol{\delta}} + \left(\mathbf{Q}^T\mathbf{D}\mathbf{Q}\right)\dot{\boldsymbol{\delta}} + \left(\mathbf{Q}^T\mathbf{K}\mathbf{Q}\right)\boldsymbol{\delta} = \mathbf{Q}^T\mathbf{T} \tag{6.19}$$

or

$$\mathsf{J}\ddot{\boldsymbol{\delta}} + \mathsf{D}\dot{\boldsymbol{\delta}} + \mathsf{K}\boldsymbol{\delta} = \mathsf{T} \tag{6.20}$$

where[1]

J = The Modal Moment of Inertia diagonal matrix
D = The Modal Damping matrix
K = The Modal Spring Constant diagonal matrix.

Usually the damping is neglected in constructing the **Q** matrix and hence the matrix D is, in general, not a diagonal matrix. In some cases, **Q** is calculated assuming that $\mathbf{D} = \mathbf{0}$, in which case $\mathsf{D} = \mathbf{0}$ as well.

---

[1]Note carefully the very plain, bold Helvetica typeface used for the transformed coefficient matrices. These transformations are defined in (6.19). The Helvetica typeface will also be used to represent the elements of these matrices, but using italic Helvetica.

Equation (6.20) represents a decoupled set of second order equations and can be solved by conventional means. The ith row of (6.20) is given by

$$J_i\ddot{\delta}_i + D_i\dot{\delta}_i + K_i\delta_i = T_i \tag{6.21}$$

and represents the modal equation for the ith mode. Usually, we write

$$J_i = 2H_i \tag{6.22}$$

where H_i is the inertia constant discussed in Chapter 4, but relates here to the moment of inertia of the ith mode only. Rewriting (6.21) in terms of the inertia constant, we have

$$2H_i\ddot{\delta}_i + D_i\dot{\delta}_i + K_i\delta_i = T_i \tag{6.23}$$

or

$$\ddot{\delta}_i + \frac{D_i}{2H_i}\dot{\delta}_i + \frac{K_i}{2H_i}\delta_i = \frac{T_i}{2H_i} \tag{6.24}$$

Usually, we write (6.24) in a "standard form" as follows.

$$\ddot{\delta}_i + 2\sigma_i\dot{\delta}_i + \omega_i^2\delta_i = \frac{T_i}{2H_i} \tag{6.25}$$

where

$$\sigma_i = \frac{T_i}{4H_i} = \text{The Modal Damping} \tag{6.26}$$

$$\omega_i = \sqrt{\frac{K_i}{2H_i}} = \text{The Modal Natural Frequency} \tag{6.27}$$

and where we note that both σ_i and ω_i can be measured in the field.

If the field measurements are available, then the spring-mass model of (6.15) can be adjusted to yield the field measured values. To do this we must refer to (6.21) and evaluate the coefficients.

$$
\begin{aligned}
J_i &= \text{The } i\text{th diagonal element of } \mathbf{Q}^t\mathbf{J}\mathbf{Q} \\
D_i &= \text{The } i\text{th diagonal element of } \mathbf{Q}^t\mathbf{D}\mathbf{Q} \\
K_i &= \text{The } i\text{th diagonal element of } \mathbf{Q}^t\mathbf{K}\mathbf{Q}.
\end{aligned} \tag{6.28}
$$

If $\mathbf{Q}$ is given by

$$
\mathbf{Q} = \begin{bmatrix} Q_{11} & Q_{12} & \cdots & Q_{1n} \\ Q_{21} & Q_{22} & \cdots & Q_{2n} \\ \cdots & \cdots & \cdots & \cdots \\ Q_{n1} & Q_{n2} & \cdots & Q_{nn} \end{bmatrix} \tag{6.29}
$$

then it can easily be shown that the diagonal terms of the transformations (6.28) are given by

$$
\begin{aligned}
D_i &= D_1 Q_{1i}^2 + D_2 Q_{2i}^2 + \ldots + D_n Q_{ni}^2 \\
J_i &= J_1 Q_{1i}^2 + J_2 Q_{2i}^2 + \ldots + J_n Q_{ni}^2.
\end{aligned} \tag{6.30}
$$

The off-diagonal terms of the $\boldsymbol{D}$ -matrix may be nonzero, but these off-diagonal terms are ignored in the model adjustment process.

Repeating for $n = 1$ to n the result can be written in matrix form as follows.

$$
\begin{bmatrix} Q_{11}^2 & Q_{21}^2 & \cdots & Q_{g1}^2 & \cdots & Q_{n1}^2 \\ Q_{12}^2 & Q_{22}^2 & \cdots & Q_{g2}^2 & \cdots & Q_{n2}^2 \\ \cdots & \cdots & \cdots & \cdots & \cdots & \cdots \\ Q_{1n}^2 & Q_{2n}^2 & \cdots & Q_{gn}^2 & \cdots & Q_{nn}^2 \end{bmatrix} \begin{bmatrix} D_1 \\ D_2 \\ \cdots \\ D_n \end{bmatrix} = \begin{bmatrix} D_1 \\ D_2 \\ \cdots \\ D_n \end{bmatrix} \tag{6.31}
$$

and

$$
\begin{bmatrix} Q_{11}^2 & Q_{21}^2 & \cdots & Q_{g1}^2 & \cdots & Q_{n1}^2 \\ Q_{12}^2 & Q_{22}^2 & \cdots & Q_{g2}^2 & \cdots & Q_{n2}^2 \\ \cdots & \cdots & \cdots & \cdots & \cdots & \cdots \\ Q_{1n}^2 & Q_{2n}^2 & \cdots & Q_{gn}^2 & \cdots & Q_{nn}^2 \end{bmatrix} \begin{bmatrix} J_1 \\ J_2 \\ \cdots \\ J_n \end{bmatrix} = \begin{bmatrix} J_1 \\ J_2 \\ \cdots \\ J_n \end{bmatrix}. \tag{6.32}
$$

6.1.2.2 Model Adjustment for Damping

To adjust the spring-mass model of (6.15) for damping, (6.31) can be solved directly to give the values of the n dashpot dampings D, given the modal dampings on the right-hand side of (6.31). However, for an n mass model, only n–1 dampings are available because one of the modes is the rigid-body mode. This problem can be overcome by arbitrarily setting the modal damping corresponding to the rigid-body mode to zero. If we assume that the nth mode corresponds to the rigid-body mode, then the dashpot dampings can be computed as follows:

$$\begin{bmatrix} Q_{11}^2 & Q_{21}^2 & \cdots & Q_{g1}^2 & \cdots & Q_{n1}^2 \\ Q_{12}^2 & Q_{22}^2 & \cdots & Q_{g2}^2 & \cdots & Q_{n2}^2 \\ \cdots & \cdots & \cdots & \cdots & \cdots & \cdots \\ Q_{1n}^2 & Q_{2n}^2 & \cdots & Q_{gn}^2 & \cdots & Q_{nn}^2 \end{bmatrix} \begin{bmatrix} D_1 \\ D_2 \\ \cdots \\ D_n \end{bmatrix} = \begin{bmatrix} D_1 \\ D_2 \\ \cdots \\ 0 \end{bmatrix}. \tag{6.33}$$

This equation can be solved directly to obtain a unique solution for the dashpot dampings. However, some of the dampings may come out to be negative, which are clearly fictitious values. Even so, this can be handled mathematically without any problem.

Another practical problem that may arise is that only a few of the modes may be easily excitable and thus the modal damping may not be known for all modes. This can also be solved by setting the modal damping arbitrarily to zero for those modes with unknown modal dampings. The relationship (6.33) will always have a unique solution as long as there is at least one mode with nonzero damping.

It should be noted that (6.33) is a solution for only the diagonal elements of the **D** matrix. The non-diagonal elements of the **D** matrix represent cross damping among the masses and are fictitious. Also, the spring-mass model of Figure 6.1 does not have any provision for representing these off-diagonal damping terms except for the damping between adjacent masses. Thus, all non-diagonal terms of the **D** matrix are set to zero. Usually this produces negligible error for modal damping values in the practical range.

A few words of caution are in order. The solution to (7.33) results in good results for low order models. For higher order models the damping values could be large (fictitious) enough to impact the frequencies.

If a complete solution for the **D**-matrix is desired it can be obtained by referring to (6.19), where

$$\mathsf{D} = \mathbf{Q}^{\mathrm{T}}\mathbf{D}\mathbf{Q}.$$

Then

$$\mathbf{D} = \left[\mathbf{Q}^{\mathrm{T}}\right]^{-1} \mathsf{D}\mathbf{Q}^{-1}. \tag{6.34}$$

From field measurements of modal damping

$$\mathsf{D} = 2\mathbf{J}\boldsymbol{\sigma} \tag{6.35}$$

where $\boldsymbol{\sigma}$ is a matrix of damping coefficients defined by (6.26). Hence,

$$\mathbf{D} = \left[\mathbf{Q}^{\mathrm{T}}\right]^{-1} (2\mathbf{J}\boldsymbol{\sigma})\mathbf{Q}^{-1}. \tag{6.36}$$

For a special case of equal damping in all modes, $\boldsymbol{\sigma}$ is a constant times the identity matrix. In this special case

$$\mathbf{D} = 2\sigma\left[\mathbf{Q}^{\mathrm{T}}\right]^{-1} \mathsf{J}\mathbf{Q}^{-1} = 2\sigma\mathsf{J}. \tag{6.37}$$

Thus the **D** matrix is a constant multiple of the **J** matrix and is diagonal. In other words the damping terms are a fixed multiple of inertia terms. The above is an exact solution and no approximations are involved. In this special case both the **D** matrix and the D matrix are diagonal.

6.1.2.3 Model Adjustment for Frequencies

If measured frequencies are found to be different from the computed values, either the model inertias or the model spring constants can be adjusted to correct the frequencies. In the following, the inertia adjustment method is outlined.

From (6.27), the modal frequency is given by

$$\omega_i^2 = \frac{K_i}{2H_i} = \frac{K_i}{J_i} \tag{6.38}$$

or

$$K_i = J_i\omega_i^2 = \text{a constant} \tag{6.39}$$

where we assume that the modal spring constant is a constant. Taking the total differential we compute

$$dK_i = d\left(J_i\omega_i^2\right) = 0 \tag{6.40}$$

or

$$\omega_i^2 dJ_i + 2\omega_i J_i d\omega_i = 0. \tag{6.41}$$

For small changes in the parameters, we can write

$$\Delta J_i = -2J_i \frac{\Delta\omega_i}{\omega_i}. \tag{6.42}$$

Thus, if there are p modes to be adjusted, then p of the terms like (6.42) will be nonzero. Now (6.32) can be rewritten in incremental form as

$$\begin{bmatrix} Q_{11}^2 & Q_{21}^2 & \cdots & Q_{g1}^2 & \cdots & Q_{n1}^2 \\ Q_{12}^2 & Q_{22}^2 & \cdots & Q_{g2}^2 & \cdots & Q_{n2}^2 \\ \cdots & \cdots & \cdots & \cdots & \cdots & \cdots \\ Q_{1n}^2 & Q_{2n}^2 & \cdots & Q_{gn}^2 & \cdots & Q_{nn}^2 \end{bmatrix} \begin{bmatrix} \Delta J_1 \\ \Delta J_2 \\ \cdots \\ \Delta J_n \end{bmatrix} = \begin{bmatrix} \Delta J_1 \\ \Delta J_2 \\ \cdots \\ \Delta J_n \end{bmatrix}. \tag{6.43}$$

Of the n equations, only p of them, corresponding to the p measured modal frequencies, are known. Once again, this problem is solved by arbitrarily setting the right-hand side of (6.43) to zero for the remaining $n–p$ modes. Then (6.43) can be solved. It is important that the modal inertia corresponding to the j th mode frequency appear in the j th row of (6.43).

6.1.2.4 Iterative Solution of the Inertia Adjustment Equations

For most practical cases, solution of (6.43) will result in small enough changes (ΔJ's) that will not affect the Q and D values appreciably. However, if greater accuracy is desired, the following iterative procedure can be used until the desired accuracy is achieved.

1. Calculate the ΔJ_k from the $\Delta\omega_k$ using (6.42) for each k.
2. Calculate ΔJ_i using (6.43) and then compute new values of J_k: $J_{k(new)} = J_{k(old)} + \Delta J_k$
3. Recalculate the **Q** matrix from the adjusted model assuming the damping D_i to be zero.
4. Recalculate J_k using (6.30).
5. Repeat steps 1-4 until the desired accuracy is achieved.
6. Verify the model by calculating ω_{ii} and σ_i using (6.26) and (6.27).

7. After all adjustments for frequencies have been made, adjust for damping using (6.33).

In the above procedure, the adjustments for the damping has been intentionally kept outside the iteration loop to keep the **Q** matrix real. Usually the dampings are small and have negligible impact on the mode shapes. Then, excluding the damping adjustments from the iterative process is justified. This is illustrated by an example.

Example 6.2 Model Adjustment for Frequency Matching

The model adjustment process is illustrated by the following simple example. A turbine-generator system is represented by the three mass model shown in Figure 6.3.

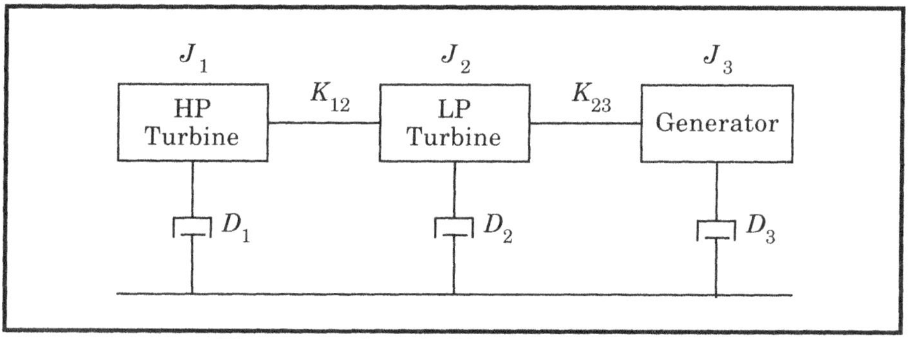

Figure 6.3 Sample Three-Mass System

The inertias and spring constants are given as

$$J_1 = 1216 \text{ lbf - ft - s}^2 \quad K_{12} = 35.28 \times 10^6 \text{ lbf} - \text{ft}$$

$$J_2 = 6975 \text{ lbf - ft - s}^2 \quad K_{23} = 70.40 \times 10^6 \text{ lbf} - \text{ft.}$$

$$J_3 = 4060 \text{ lbf - ft - s}^2$$

Then the dynamics of the above spring-mass model system are given by

$$\begin{bmatrix} 1216.0 & & \\ & 6975.0 & \\ & & 4060.0 \end{bmatrix} \begin{bmatrix} \ddot{\theta}_1 \\ \ddot{\theta}_2 \\ \ddot{\theta}_3 \end{bmatrix} + \begin{bmatrix} D_1 & & \\ & D_2 & \\ & & D_3 \end{bmatrix} \begin{bmatrix} \dot{\theta}_1 \\ \dot{\theta}_2 \\ \dot{\theta}_3 \end{bmatrix}$$
$$+ 10^6 \begin{bmatrix} +35.28 & -35.28 & 0 \\ -35.28 & +105.70 & -70.40 \\ 0 & -70.40 & +70.40 \end{bmatrix} \begin{bmatrix} \theta_1 \\ \theta_2 \\ \theta_3 \end{bmatrix} = \begin{bmatrix} T_1 \\ T_2 \\ T_3 \end{bmatrix} \tag{6.44}$$

The above is a coupled system of equations. The transformation matrix that will decouple this system of equations is found by computing the mode shapes (eigenvectors) of the above system. The damping constants are unknown, but they are usually small and can be assumed zero for the computation of the eigenvectors. The modal frequencies and mode shapes for the above system are computed as follows. First, the eigenvalues are found to be

$$\begin{aligned} \lambda_1 &= 0 + j196.53 \text{ s}^{-1} \\ \lambda_2 &= 0 + j151.24 \text{ s}^{-1} \\ \lambda_3 &= 0 + j0. \end{aligned} \tag{6.45}$$

Using the eigenvalues, we compute the modes shapes (eigenvectors) as shown in Table 6.1.

Note the following concerning the system mode shapes:

1. Each mode shape corresponds to a specific eigenvalue from (6.45). Thus the numbering of eigenvalues in (6.45) is arbitrary but once the eigenvalues have been numbered, the mode shapes in Table 6.1 must correspond to the eigenvalue numbers.

2. All mode shapes have been normalized with respect to the generator mass.

3. Mode 3 is the rigid-body mode. In this mode all three masses move exactly together.

The modal frequencies are computed from the eigenvalues by dividing the imaginary part by 2π. Thus, we get the following frequencies.

Table 6.1 Sample System Eigenvectors				
Masses		Mode Shapes		
Mass Name	Mass No.	Mode 1	Mode 2	Mode 3
HP	1	3.7020	-1.5085	1.0000
LP	2	-1.2270	-0.3192	1.0000
Gen	3	1.0000	1.0000	1.0000

$$\begin{aligned} \omega_1 &= 196.53 \text{ rad/s} \qquad f_1 = 31.29 \text{ Hz} \\ \omega_2 &= 151.24 \text{ rad/s} \qquad f_2 = 24.08 \text{ Hz} \\ \omega_3 &= 0.00 \text{ rad/s} \qquad f_3 = 0.00 \text{ Hz} \end{aligned} \tag{6.46}$$

The transformation matrix $\mathbf{Q}$ of (6.16) is constructed by inserting the eigenvectors in the rows of the matrix. Thus we construct the following matrix.

$$\mathbf{Q} = \begin{bmatrix} +3.7020 & -1.5085 & +1.0000 \\ -1.2270 & -0.3192 & +1.0000 \\ +1.0000 & +1.0000 & +1.0000 \end{bmatrix} \tag{6.47}$$

The modal inertias are then computed using (6.30) as follows.

$$\begin{aligned} J_1 &= J_1 Q_{11}^2 + J_2 Q_{21}^2 + J_3 Q_{31}^2 \\ &= (1216.0)(3.7020)^2 + (6976.0)(1.2270)^2 + (4060.0)(1.0)^2 \\ &= 31{,}226 \text{ lbf-ft-s}^2 \end{aligned} \tag{6.48}$$

$$\begin{aligned} J_2 &= J_1 Q_{12}^2 + J_2 Q_{22}^2 + J_3 Q_{32}^2 \\ &= (1216.0)(-1.5085)^2 + (6976.0)(-0.3192)^2 + (4060.0)(1.0)^2 \\ &= 7{,}537 \text{ lbf-ft-s}^2 \end{aligned} \tag{6.49}$$

$$\begin{aligned} J_3 &= J_1Q_{13}^2 + J_2Q_{23}^2 + J_3Q_{33}^2 \\ &= (1216.0)(1.0)^2 + (6976.0)(1.0)^2 + (4060.0)(1.0)^2 \\ &= 12{,}251 \text{ lbf-ft-s}^2. \end{aligned} \tag{6.50}$$

Thus we write

$$\mathbf{J} = \begin{bmatrix} 31226 & 0 & 0 \\ 0 & 7537 & 0 \\ 0 & 0 & 12251 \end{bmatrix}. \tag{6.51}$$

Now, let us assume that we need to adjust the model to obtain the following frequencies and dampings, which have been obtained from field measurements.

$$\begin{aligned} f_1 &= 31.40 \text{ Hz} \qquad \sigma_1 = 0.10 \\ f_2 &= 23.95 \text{ Hz} \qquad \sigma_2 = 0.20. \end{aligned} \tag{6.52}$$

First, we adjust the model for frequency. From (6.42) we write

$$\Delta J_i = -2J_i \frac{\Delta\omega_i}{\omega_i} = -2J_i \frac{\Delta f_i}{f_i}. \tag{6.53}$$

Hence we compute

$$\begin{aligned} \Delta J_1 &= -2(31{,}226)\frac{(31.40 - 31.29)}{31.29} = -219.55 \\ \Delta J_2 &= -2(7{,}537)\frac{(23.95 - 24.08)}{24.08} = +81.38 \\ \Delta J_3 &= 0. \end{aligned} \tag{6.54}$$

Since the Mode 3 frequency does not require adjustment, we assume the third inertia adjustment to be zero. Substituting (6.47) into (6.43) we get

$$\begin{bmatrix} 3.7020^2 & 1.2270^2 & 1.0^2 \\ 1.5085^2 & 0.3192^2 & 1.0^2 \\ 1.0^2 & 1.0^2 & 1.0^2 \end{bmatrix} \begin{bmatrix} \Delta J_1 \\ \Delta J_2 \\ \Delta J_3 \end{bmatrix} = \begin{bmatrix} \Delta J_1 \\ \Delta J_2 \\ \Delta J_3 \end{bmatrix}. \tag{6.55}$$

Substituting the values from (6.54) we get

$$\begin{bmatrix} 3.7020^2 & 1.2270^2 & 1.0^2 \\ 1.5085^2 & 0.3192^2 & 1.0^2 \\ 1.0^2 & 1.0^2 & 1.0^2 \end{bmatrix} \begin{bmatrix} \Delta J_1 \\ \Delta J_2 \\ \Delta J_3 \end{bmatrix} = \begin{bmatrix} -219.55 \\ +81.38 \\ 0.00 \end{bmatrix}. \tag{6.56}$$

This equation can be solved for the changes in the inertias to get

$$\begin{bmatrix} \Delta J_1 \\ \Delta J_2 \\ \Delta J_3 \end{bmatrix} = \begin{bmatrix} -12.951 \\ -108.887 \\ +42.668 \end{bmatrix}. \tag{6.57}$$

We apply these changes to the original model in (6.44) to get:

$$\begin{bmatrix} 1203.05 & & \\ & 6866.11 & \\ & & 4102.67 \end{bmatrix} \begin{bmatrix} \ddot{\theta}_1 \\ \ddot{\theta}_2 \\ \ddot{\theta}_3 \end{bmatrix} + \begin{bmatrix} D_1 & & \\ & D_2 & \\ & & D_3 \end{bmatrix} \begin{bmatrix} \dot{\theta}_1 \\ \dot{\theta}_2 \\ \dot{\theta}_3 \end{bmatrix}$$

$$+ 10^6 \begin{bmatrix} +35.28 & -35.28 & 0 \\ -35.28 & +105.70 & -70.40 \\ 0 & -70.40 & +70.40 \end{bmatrix} \begin{bmatrix} \theta_1 \\ \theta_2 \\ \theta_3 \end{bmatrix} = \begin{bmatrix} T_1 \\ T_2 \\ T_3 \end{bmatrix}. \tag{6.58}$$

Note that the change alters only the inertia matrix. Usually, for small changes in frequencies, one iteration is adequate. However, if greater accuracy is desired, the process from (6.44) to (6.57) can be repeated until the desired accuracy is obtained.

We now adjust our model for the dampings. For this case we have damping known for the first two modes, and these dampings are in units of radians per second. We first convert these dampings to the proper units as follows.

$$D_i = 4H_i\sigma_i = 2J_i\sigma_i \tag{6.59}$$

Usually, one would use (6.58) to compute new modal inertias and new mode shapes for determining the dampings. For the sake of simplicity, we assume that there is no change in modal frequencies and hence use (6.44) to compute the Ds, rather than the updated model of (6.58). This usually causes no significant error. Substituting for the modal inertias from (6.51) we get

$$\begin{aligned} D_1 &= 2(31{,}226)(0.10) = 6245.2 \\ D_2 &= 2(7{,}537).(0.20) \ = 3014.8 \\ D_3 &= 0. \end{aligned} \tag{6.60}$$

As before, we set the damping of the third mode to zero. These values are substituted into (6.33) to obtain the following dashpot dampings:

$$\begin{bmatrix} 3.7020^2 & 1.2270^2 & 1.0^2 \\ 1.5085^2 & 0.3192^2 & 1.0^2 \\ 1.0^2 & 1.0^2 & 1.0^2 \end{bmatrix} \begin{bmatrix} D_1 \\ D_2 \\ D_3 \end{bmatrix} = \begin{bmatrix} 6245.2 \\ 3014.8 \\ 0.0 \end{bmatrix}. \tag{6.61}$$

Solving, we get

$$\begin{bmatrix} D_1 \\ D_2 \\ D_3 \end{bmatrix} = \begin{bmatrix} 591.7 \\ -2516.1 \\ 1924.4 \end{bmatrix}. \tag{6.62}$$

Substituting these values of D into (6.44) gives the desired corrections to the model to account for the measured dampings. Since the inertia adjustments for frequencies are small, (6.62) can be directly substituted into (6.58) to give the following model, which will include corrections for both the frequencies and the dampings:

$$\begin{bmatrix} 1203.05 & & \\ & 6866.11 & \\ & & 4102.67 \end{bmatrix} \begin{bmatrix} \ddot{\theta}_1 \\ \ddot{\theta}_2 \\ \ddot{\theta}_3 \end{bmatrix} + \begin{bmatrix} 591.7 & & \\ & -2516.1 & \\ & & 1924.1 \end{bmatrix} \begin{bmatrix} \dot{\theta}_1 \\ \dot{\theta}_2 \\ \dot{\theta}_3 \end{bmatrix}$$

$$+\,10^6 \begin{bmatrix} +35.28 & -35.28 & 0 \\ -35.28 & +105.70 & -70.40 \\ 0 & -70.40 & +70.40 \end{bmatrix} \begin{bmatrix} \theta_1 \\ \theta_2 \\ \theta_3 \end{bmatrix} = \begin{bmatrix} T_1 \\ T_2 \\ T_3 \end{bmatrix}. \tag{6.63}$$

Note that one of the dampings is negative, which has no physical meaning, but this causes no problem for the eigenvalue program.

Solving (6.63) for eigenvalues gives the following values of modal frequencies and dampings:

$$f_1 = 31.43 \text{ Hz} \qquad \sigma_1 = 0.1003 \text{ s}^{-1}$$
$$f_2 = 24.07 \text{ Hz} \qquad \sigma_2 = 0.1969 \text{ s}^{-1}. \tag{6.64}$$

This compares with the desired values of (6.52), repeated here for comparison.

$$f_1 = 31.40 \text{ Hz} \qquad \sigma_1 = 0.10 \text{ s}^{-1}$$
$$f_2 = 23.95 \text{ Hz} \qquad \sigma_2 = 0.20 \text{ s}^{-1}. \tag{6.65}$$

It is seen that the adjusted model of (6.63) accurately predicts the damping but the frequencies are not exactly right. However, it achieved approximately 50% of the desired frequency adjustment. If more accurate frequency predictions are desired, the model adjustment process of (6.44) to (6.58) can be repeated until the desired accuracy is achieved.

This completes Example 6.2.

As long as the damping adjustments of the model are relatively small, they can be treated as being independent of the frequency adjustments. Thus, the mode shapes can be calculated with dampings neglected. This mode shape information is needed in order to calculate subsequent adjustments to the model.

6.2 THE MODAL MODEL

An alternative to the damping representation discussed above is to represent the turbine-generator system dynamics in modal form. The analytical method employed, such as an eigenvalue method, can be arranged to provide an option to directly input the modal model parameters of (6.20) and the **Q**-matrix.

In a real system the coupling between the turbine-generator and the electrical system occurs only through generator mass. Since θ_g, $p\theta_g$ (where p is the derivative with respect to time), and T_g are the only variables needed to calculate coupling, only the row of the **Q** matrix which corresponds to the generator mass needs to be known. Thus, one only needs to know the vector θ_g, which is defined as follows.

$$\boldsymbol{\theta}_g = \begin{bmatrix} \theta_{g1} & \theta_{g2} & \cdots & \theta_{gn} \end{bmatrix}^{\mathrm{T}}. \tag{6.67}$$

Knowing this vector, the modal inertia, the modal damping, and the modal spring constants of (6.23) one may calculate all the eigenvalues of interest.

One advantage of the modal model representation is that a high order spring-mass system can be replaced by a lower order modal model where only the subsynchronous modes are represented. Generally, the error introduced by neglecting the effect of higher order modes is negligible.

6.3 FIELD TESTS FOR FREQUENCIES AND DAMPING

With the advent of digital signal analysis (DSA) and Fast Fourier Transform (FFT) techniques, the measurement of the torsional mechanical frequencies of a turbine-generator is relatively easy. Generally, the rotor is equipped with a toothed wheel and with pickups to measure the shaft speed deviation. The output of the pickup is a frequency modulated (FM) signal containing the velocity deviation information. The test arrangement is shown in Figure 6.4. The speed signal from the pickup is processed by an FM demodulator to produce the velocity deviation signal.

Under normal conditions the unit is operating at some steady-state operating point. Even under steady-state conditions on the actual power system, the torque on the turbine-generator rotor is never absolutely constant. The effect of these small torque deviations is to cause a natural ringing of the rotor at the natural frequencies of the shaft. This information is picked up by the toothed wheel and is processed by the FM demodulator. The output of the demodulator, if plotted as a function of time, would look like a random noise signal. However, this signal contains the frequency information of each natural oscillatory mode of the shaft. To determine these natural frequencies, the signal is fed into a DSA, which converts the time domain information into the frequency domain. The autospectrum of the signal contains the modal frequency information, but it

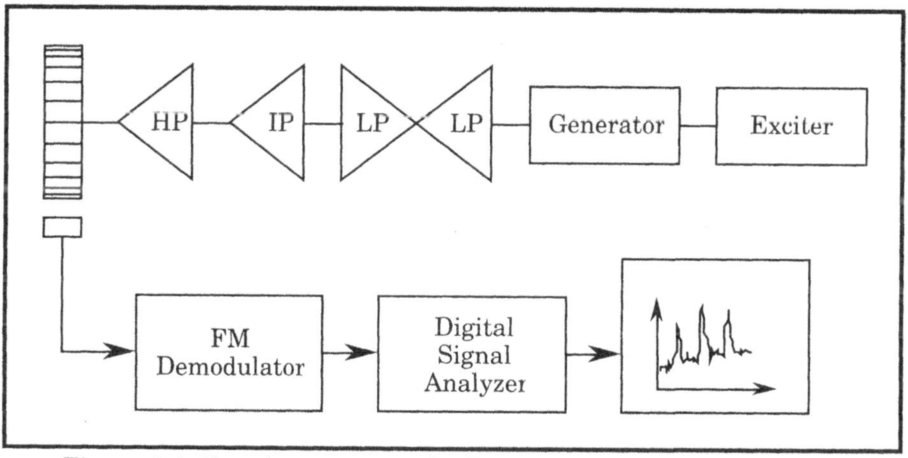

Figure 6.4 Test Arrangement for Determining Shaft Parameters

is generally buried in the background noise. The DSA makes repeated measurements and uses an averaging technique to average out the noise and retain the torsional frequency information. After only a few minutes of such averaging, the modal frequencies can be distinctly recognized. A typical DSA output, taken from an actual measurement, is shown in Figure 6.5. These plots show that the unit tested has five modes of oscillation. Two modes are subsynchronous, occurring at 24.1 and 31.2 Hz. Two modes are supersynchronous, occurring at 91.0, 111.2 Hz. The peaks shown at 60 Hz and 120 Hz in both plots are due to system noise. The 120 Hz peak appears to be the larger of the two since the measurement system has a 60 Hz filter, which attenuates that frequency.

6.4 DAMPING TESTS

The measurement of modal damping is much more complicated than measuring the frequency alone. It requires that a means be provided for exciting the torsionals and measuring the natural rate of decay corresponding to a known system condition. Two methods are commonly used to excite the torsional frequencies: a transient method and a steady-state method.

6.4.1 Transient Method

In the transient method, a system disturbance, such as a line switching or unit synchronization is used to create a sudden change in electrical torque of the unit under test. This switching causes all torsional modes to be

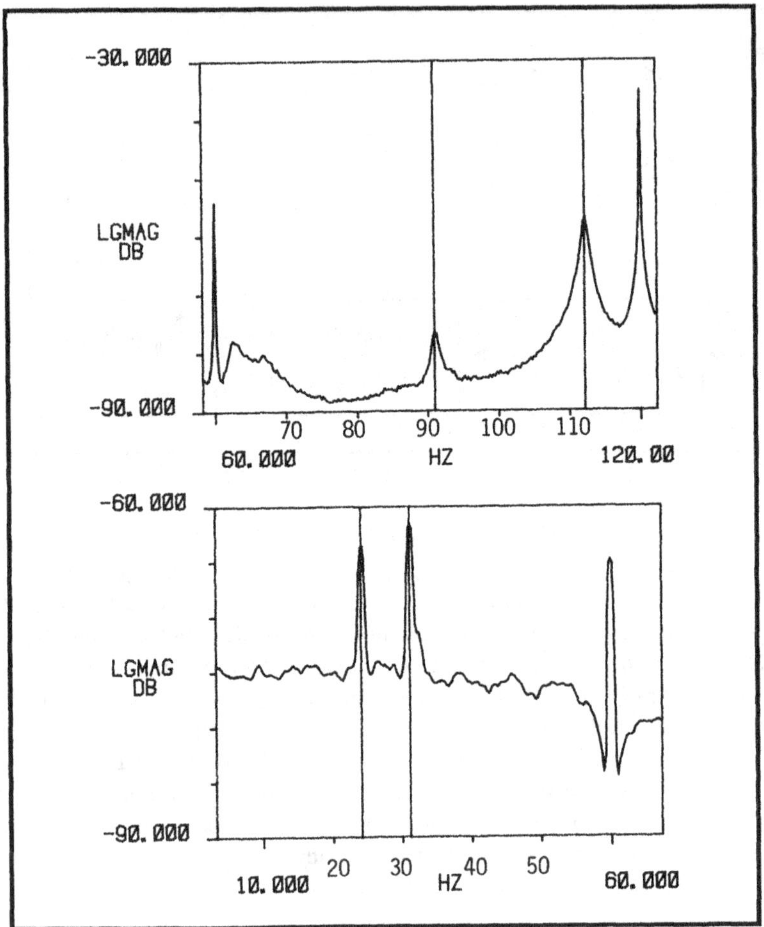

Figure 6.5 Frequency Spectra for a Turbine-Generator Unit in the Supersynchronous (top) and Subsynchronous (bottom) Frequency Ranges

excited, and consequently the $\Delta\omega$ signal contains a component of each of the modal frequencies. Each of the components reaches a peak value generally within 0.5 seconds and then decays at its natural rate.

6.4.2 Steady-State Method

The steady-state method uses a sinusoidal signal corresponding to one of the natural torsional frequencies as an excitation to the shaft system. This signal is introduced into the generator excitation system, which produces generator airgap torques, and consequently shaft torques at the selected frequency. The test signal is slowly increased until the signal reaches a predetermined level, at which time the test signal is removed. The shaft oscillation then decays at its natural rate, and these oscillations are measured to provide damping information. For some units, this method may not be feasible due to excessive attenuation in the excitation system at the higher torsional frequencies.

6.4.3 Speed Signal Processing

Obtaining the damping information from the decaying signal involves extensive signal processing. First, the signal is conditioned to reject the high frequency components. Next, the modal component must be separated by either filtering or by utilizing the DSA to perform frequency domain analysis. In practice, both methods are used and the results are compared as a means of validation of the calculations.

Usually the foregoing test must be repeated several times in order to reduce the data spread and to find the damping as a function of loading and other system operating conditions.

6.4.4 Other Methods

Some turbine-generators are not equipped with toothed wheels for providing the shaft speed deviation information. Retrofitting these units with toothed wheels could be very costly and time consuming. One way to obtain the frequency and damping measurements without the use of a toothed wheel is to monitor the generator torque following a significant system transient (staged or natural). The disadvantage of this method is that all modes are not equally excited. Also, the staged tests are costly and time consuming. This method has been successfully tested by one of the authors on one machine, but needs to be validated on other units before complete success can be claimed.

6.4.5 Other Factors

Shaft torsional damping is known to be a function of load on the turbine. The spring-mass model spring constants are a property of the shaft material alone and these parameters do not change with loading. The damping parameters, on the other hand, are known to vary with the steam loading on the turbine blading. Some small contribution to damping is likely due to the shaft material, but steam damping is often considered the predominant factor. This means that any measurement of damping must be repeated at different unit loading conditions. This may involve system reconfiguration and unit rescheduling, and may require that the damping measurements be performed during light load seasons, on weekends, or late at night when the unit can be rescheduled.

6.5 REFERENCES FOR CHAPTER 6

1. Agrawal, B. L., and R. G. Farmer, "Subsynchronous Resonance Analysis and Control," Short Course Notes, Arizona State University, Winter Power Institute, Dec. 7-9, 1982.

2. Ramey, D. G., J. A. Demcko, R. G. Farmer and B. L. Agrawal, "Subsynchronous Resonance Tests and Torsional Monitoring System Verification at the Cholla Station," **IEEE Trans.**, v. PAS-99, n. 5, Sept-Oct 1980, p. 1900-1907.

3. IEEE Subsynchronous Resonance Working Group, "First Benchmark Model for Computer Simulation of Subsynchronous Resonance," ibid, v. PAS-99, September/October 1980.

4. Walker, D. N., C. E. J. Bowler, R. L. Jackson and D. A. Hodges, "Results of Subsynchronous Resonance Test at Mohave," ibid, v. PAS-94, n. 5, Sept-Oct 1975, p. 1878-1889.

5. Walker, D. N., and A. L. Schwalb, "Results of Subsynchronous Resonance Test at Navajo," IEEE Special Publication 76 CH 1066-0-PWR, p. 37-45.

6. Ramey, D. G., P. F. Harrold, H. A. Maddox, R. B. Starnes and J. L. Knickerbocker, **Trans. of ASME, Engineering for Power**, v. 99, n. 3, July 1977, p. 378-384.

7. Walker, D. N., C. E. J. Bowler, and D. H. Baker, "Torsional Dynamics of Closely Coupled Turbine Generators," **IEEE Trans.**, v. PAS-97, n. 4, July-Aug 1978, p. 1458-1465.

8. Agrawal, B. L, and R. G. Farmer, "Effective Damping for SSR Analysis of Parallel Turbine Generators," ibid, Nov. 1988, p. 1441-1448.

Part 4

System Analysis

Eigen Analysis

Eigenvalue Analysis of Benchmark Systems

CHAPTER 7

EIGEN ANALYSIS

This monograph describes a method of finding the eigenvalues and eigenvectors of the linearized model of the power system for the analysis of SSR problems. The previous sections have been concerned with the formation of the model. In this section we will describe the method used to compute the desired eigenvalues and eigenvectors from the model equations and to interpret the results.

7.1 STATE-SPACE FORM OF SYSTEM EQUATIONS

The state-space form of the system equations is nothing more than the differential equations describing the system written as a set of first-order simultaneous equations in matrix form. In control system literature, these are often written as

$$\begin{aligned} \dot{\mathbf{Y}}_1 &= \mathbf{A}\mathbf{Y}_1 + \mathbf{B}\mathbf{F} \\ \mathbf{Y}_2 &= \mathbf{C}\mathbf{Y}_1 + \mathbf{D}\mathbf{F} \end{aligned} \tag{7.1}$$

where **A**, **B**, **C** and **D** are matrices of appropriate dimensions, **F** is a vector of input variables, $\mathbf{Y}_1$ is the vector of *state variables*, and $\mathbf{Y}_2$ is a vector of output variables. This is the same form as equation 3.1.

Note that the first equation of (7.1) is a set of differential equations, whereas the second equation is algebraic. Usually, we say that the number of differential equations is n, but the number of algebraic equations can be anything the user desires. Thus the matrix **A** is always $n \times n$, but the matrix **C** need not even be square, and will always have n columns.

The choice as to which variables are included as state variables in $\mathbf{Y}_1$ is not unique. For example, there are several practical choices among voltages, currents, and flux linkages that can be made for state variables to describe a synchronous machine. But the number of state variables will be the same in all cases and is equal to the order of the model, and the value of the eigenvalues will be independent of the choice made.

We will not be considering the effect of an input forcing function and thus can ignore the vector **F**. If the state variables are chosen to adequately describe the system, there will be no need for the $\mathbf{Y}_2$ vector, and equation (7.1) can be reduced to the form

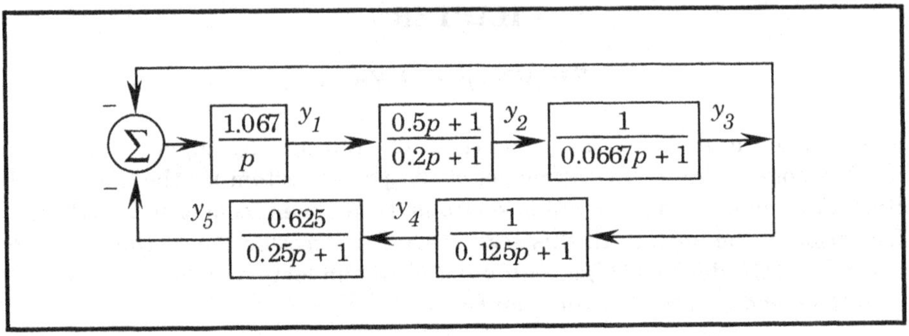

Figure 7.1 Block Diagram of the Sample Dynamic System

$$\dot{\mathbf{Y}} = \mathbf{AY}. \tag{7.2}$$

The elements of the n element **Y** vector will be the voltages, currents, flux linkages, angular velocities, etc., necessary to describe the system and the matrix **A** is an $n \times n$ matrix. The variables **Y** are called the *state variables* because, if all these values are known at any one instant in time, the system "state" is defined at that instant. For example, the initial conditions, the value of the **Y** vector at $t = 0$, completely describe the system at the start of the solution of the differential equations.

Example 7.1 The ideas stated above can best be understood in terms of a sample system. Figure 7.1 is the block diagram of a dynamic system that is to be described by equations in the state variable form. To get these equations, first write the individual equations for each block as follows:

$$\begin{aligned}
\dot{y}_1 &= 1.067(-y_3 - y_5) \\
0.2\dot{y}_2 &= 0.5\dot{y}_1 + y_1 - y_2 \\
0.0667\dot{y}_3 &= y_2 - y_3 \\
0.125\dot{y}_4 &= y_3 - y_4 \\
0.25\dot{y}_5 &= 0.625y_4 - y_5.
\end{aligned} \tag{7.3}$$

To put the equations into the form of (7.2), each equation must be divided by the coefficient on the left side of the equal sign, and the first equation must be substituted into the right side of the second equation to eliminate the derivative term of y_1. The resulting equations can be written as follows:

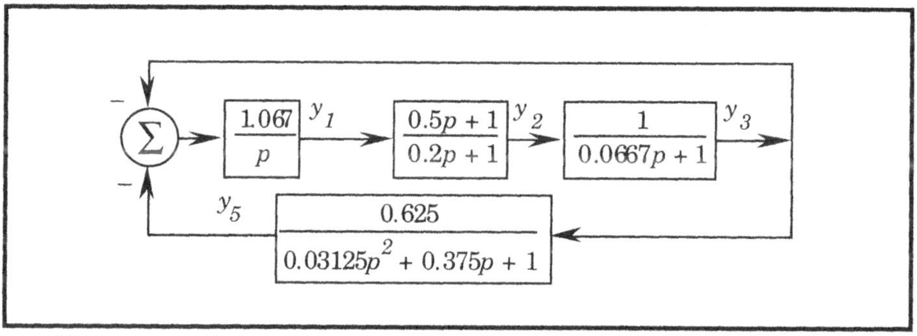

Figure 7.2 Alternate Form of the Dynamic System Block Diagram

$$\begin{bmatrix} \dot{y}_1 \\ \dot{y}_2 \\ \dot{y}_3 \\ y_4 \\ \dot{y}_5 \end{bmatrix} = \begin{bmatrix} & & -1.067 & & -1.067 \\ 5.0 & -5.0 & -2.667 & & -2.667 \\ & 15.0 & -15.0 & & \\ & & 8.0 & -8.0 & \\ & & & 2.5 & -4.0 \end{bmatrix} \begin{bmatrix} y_1 \\ y_2 \\ y_3 \\ y_4 \\ y_5 \end{bmatrix}. \tag{7.4}$$

This is one form of the state variable equations for this system, where the state variables are the y's subscripted 1 through 5.

Now redraw the system of Figure 7.1 in the form of Figure 7.2.

The dynamic system is exactly the same as before; only the form has been changed. The equations for each block can be written from inspection of the block diagram as follows:

$$\begin{aligned} \dot{y}_1 &= 1.067(-y_3 - y_5) \\ 0.2\dot{y}_2 &= 0.5\dot{y}_1 + y_1 - y_2 \\ 0.0667\dot{y}_3 &= y_2 - y_3 \\ 0.03125\ddot{y}_5 + 0.375\dot{y}_5 + y_5 &= 0.625y_3. \end{aligned} \tag{7.5}$$

The first three equations in (7.5) can be handled the same as in the previous case. But equation four, which is a second order differential equation, must be separated into two first order differential equations. One way of doing this is to define a new variable v such that

$$\dot{y}_5 = \upsilon. \tag{7.6}$$

Equations (7.5) can then be written in matrix form as follows:

$$\begin{bmatrix} \dot{y}_1 \\ \dot{y}_2 \\ \dot{y}_3 \\ \dot{y}_5 \\ \dot{\upsilon} \end{bmatrix} = \begin{bmatrix} & & -1.067 & -1.067 & \\ 5.0 & -5.0 & & & \\ & 15.0 & -15.0 & & \\ & & & & 1.0 \\ & & 20.0 & -32.0 & -12.0 \end{bmatrix} \begin{bmatrix} y_1 \\ y_2 \\ y_3 \\ y_5 \\ \upsilon \end{bmatrix} \tag{7.7}$$

This is a fifth order system, since there are only five differential equations. The five state variables are those defined in (7.7). These are not the same as the state variables as defined in (7.4), but they are an equally valid description of the system. This completes Example 7.1.

The above example shows that there can be different choices of state variables that one could define, all of which could be valid state variable descriptions.

7.2 SOLUTION OF THE STATE EQUATIONS

Equation (7.2) can be solved using the standard techniques of assuming an exponential form for the solution and substituting into the equation. Assume a solution of the form

$$\mathbf{Y} = c\,\mathbf{X}e^{\lambda t} \tag{7.8}$$

where c and λ are scalar constants and $\mathbf{X}$ is a vector of constants. Taking the time derivative of (7.8) gives

$$\dot{\mathbf{Y}} = \lambda c\,\mathbf{X}e^{\lambda t}. \tag{7.9}$$

Now, substituting (7.9) and (7.8) into (7.2) gives

$$\lambda c\mathbf{X}e^{\lambda t} = \mathbf{A}c\mathbf{X}e^{\lambda t}. \tag{7.10}$$

The two scalars, c and $e^{\lambda t}$, can be cancelled from this equation and the terms rearranged to give

$$\mathbf{AX} = \lambda \mathbf{X}$$
$$(\mathbf{A} - \mathbf{U}\lambda) = 0 \tag{7.11}$$

which is the classical eigenvalue/eigenvector equation. As before, **U** is an identity matrix. Thus we see that (7.8) does satisfy (7.2) as long as λ is an eigenvalue of **A**, and **X** is the corresponding eigenvector. Since there will be as many λ , **X** pairs as the order of the system, and if it is assumed that no two λ 's are equal, the complete matrix solution will be of the form

$$\mathbf{Y}(t) = c_1 \mathbf{X}_1 e^{\lambda_1 t} + c_2 \mathbf{X}_2 e^{\lambda_2 t} + \ldots + c_n \mathbf{X}_n e^{\lambda_n t} \tag{7.12}$$

which can also be written in expanded form as

$$\mathbf{Y}(t) = \begin{bmatrix} | & | & & | \\ \mathbf{X}_1 & \mathbf{X}_2 & \cdots & \mathbf{X}_n \\ | & | & & | \end{bmatrix} \begin{bmatrix} e^{\lambda_1 t} & & & \\ & e^{\lambda_2 t} & & \\ & & \ddots & \\ & & & e^{\lambda_n t} \end{bmatrix} \begin{bmatrix} c_1 \\ c_2 \\ \cdot \\ \cdot \\ \cdot \\ c_n \end{bmatrix}. \tag{7.13}$$

Now, by setting $t = 0$, it is seen that

$$\begin{bmatrix} c_1 \\ c_2 \\ \cdot \\ \cdot \\ \cdot \\ c_n \end{bmatrix} = \begin{bmatrix} | & | & & | \\ \mathbf{X}_1 & \mathbf{X}_2 & \cdots & \mathbf{X}_n \\ | & | & & | \end{bmatrix}^{-1} \mathbf{Y}(0). \tag{7.14}$$

Using a similar notation, equation (7.11) can be rewritten for all of the λ , $\mathbf{X}$ pairs as

$$\mathbf{A}\begin{bmatrix} | & | & & | \\ \mathbf{X}_1 & \mathbf{X}_2 & \cdots & \mathbf{X}_n \\ | & | & & | \end{bmatrix} = \begin{bmatrix} | & | & & | \\ \mathbf{X}_1 & \mathbf{X}_2 & \cdots & \mathbf{X}_n \\ | & | & & | \end{bmatrix}\begin{bmatrix} \lambda_1 & & & \\ & \lambda_2 & & \\ & & \cdots & \\ & & & \lambda_n \end{bmatrix}. \tag{7.15}$$

Pre- and post-multiplying by the inverse of the matrix of eigenvectors, and using the notation

$$\begin{bmatrix} | & | & & | \\ \mathbf{X}_1 & \mathbf{X}_2 & \cdots & \mathbf{X}_n \\ | & | & & | \end{bmatrix}^{-1} = \begin{bmatrix} \text{---} \mathbf{V}_1 \text{---} \\ \text{---} \mathbf{V}_2 \text{---} \\ \cdots \\ \text{---} \mathbf{V}_n \text{---} \end{bmatrix} \tag{7.16}$$

gives

$$\begin{bmatrix} \text{---} \mathbf{V}_1 \text{---} \\ \text{---} \mathbf{V}_2 \text{---} \\ \cdots \\ \text{---} \mathbf{V}_n \text{---} \end{bmatrix}\mathbf{A} = \begin{bmatrix} \lambda_1 & & & \\ & \lambda_2 & & \\ & & \cdots & \\ & & & \lambda_n \end{bmatrix}\begin{bmatrix} \text{---} \mathbf{V}_1 \text{---} \\ \text{---} \mathbf{V}_2 \text{---} \\ \cdots \\ \text{---} \mathbf{V}_n \text{---} \end{bmatrix} \tag{7.17}$$

or

$$\mathbf{A}^{\mathrm{T}}\begin{bmatrix} | & | & & | \\ \mathbf{V}_1 & \mathbf{V}_2 & \cdots & \mathbf{V}_n \\ | & | & & | \end{bmatrix} = \begin{bmatrix} | & | & & | \\ \mathbf{V}_1 & \mathbf{V}_2 & \cdots & \mathbf{V}_n \\ | & | & & | \end{bmatrix}\begin{bmatrix} \lambda_1 & & & \\ & \lambda_2 & & \\ & & \cdots & \\ & & & \lambda_n \end{bmatrix}. \tag{7.18}$$

The **X**'s are sometimes called the right-hand eigenvectors and the **V**'s the left-hand eigenvectors or the eigenvectors of the transposed matrix. Note that (7.16) is true only if the eigenvectors are properly normalized.

Now let us consider the interpretation of (7.12). Suppose that for a given fourth order system we get the following result (not from a real system):

$$\begin{bmatrix} \psi_d \\ \psi_q \\ i_F \\ i_a \end{bmatrix} = 10.\begin{bmatrix} 100. \\ 0. \\ 50. \\ 10. \end{bmatrix} e^{-5t} + 5.\begin{bmatrix} 0. \\ 20. \\ 0. \\ 10. \end{bmatrix} e^{-8t} + (6+j5)\begin{bmatrix} 10+j5 \\ 0. \\ 3-j4 \\ 2+j1 \end{bmatrix} e^{-(5+j4)t} + (6-j5)\begin{bmatrix} 10-j5 \\ 0. \\ 3+j4 \\ 2-j1 \end{bmatrix} e^{-(5-j4)t} \tag{7.19}$$

where the exponential coefficients of t represent the four eigenvalues and the column matrices or the right-hand side of (7.19) represent the eigenvectors. The eigenvalues are the familiar modes of response for the system. They must be in the left half of the complex plane if the system is to be stable, and complex eigenvalues will occur in complex pairs representing terms of the form $e^{\alpha t}\cos(\beta t + \phi)$.

We now examine the important information conveyed by the eigenvectors. The first term to the right of (7.19) is an exponential decay due to the real eigenvalue ($\lambda = -5$). The elements of the corresponding eigenvector show that this mode of response is not present in the variable ψ_q, that it is five times larger in i_F than in i_a, and that it is 10 times larger in ψ_d than in i_a. In a similar manner, we see that the e^{-8t} response is not present in ψ_d or i_F, and that it is twice as large in ψ_q as in i_a. For the complex eigenvalues, the elements of the eigenvectors are complex. As with the real eigenvectors, the magnitude of the terms give the relative magnitude of the response of that mode in the different variables. The angle of the term gives the relative phase of this mode of response in that particular variable.

The eigenvalues and eigenvectors are functions of the system itself, and not of the initial conditions. The initial conditions come in through the c constants in (7.7). Substituting (7.16) into (7.14) gives

$$\begin{bmatrix} c_1 \\ c_2 \\ \cdots \\ c_n \end{bmatrix} = \begin{bmatrix} \text{———} \mathbf{V}_1 \text{———} \\ \text{———} \mathbf{V}_2 \text{———} \\ \cdots \\ \text{———} \mathbf{V}_n \text{———} \end{bmatrix} \begin{bmatrix} Y_1(0) \\ Y_2(0) \\ \cdots \\ Y_n(0) \end{bmatrix} . \tag{7.20}$$

Thus, it is seen that the *right hand eigenvectors* show the distribution of the modes of response through the state variables, while the *left hand eigenvectors* show the relative effect of the different initial conditions on the mode of response. The initial conditions are not as important in the SSR problem as they would be in some other problems, but their effect can be examined in a rigorous and straightforward manner.

As was seen in the last section, there are many choices that can be made for the state variables to describe a system. Since the eigenvalues are a characteristic of the system, their value will be independent of the choice made for the state variables. But the eigenvectors will change with this choice. For example, two sets of state equations were given for the previous example in (7.4) and (7.7). Each give the same five values for the eigenvalues, namely

$$\lambda_1 = -1.228$$
$$\lambda_2 \text{ and } \lambda_3 = -3.852 \pm j2.819$$
$$\lambda_4 \text{ and } \lambda_5 = -11.534 \pm j3.961$$

But, looking at the eigenvector for the real eigenvalue −1.228, from (7.4) we get

$$\begin{bmatrix} X_1 \\ X_2 \\ X_3 \\ X_4 \\ X_5 \end{bmatrix} = \begin{bmatrix} 1.000 \\ 0.512 \\ 0.557 \\ 0.658 \\ 0.594 \end{bmatrix} \tag{7.21}$$

and from equation (7.7)

$$\begin{bmatrix} X_1 \\ X_2 \\ X_3 \\ X_4 \\ V \end{bmatrix} = \begin{bmatrix} 1.000 \\ 0.512 \\ 0.557 \\ 0.594 \\ -0.729 \end{bmatrix} \tag{7.22}$$

Note that the eigenvector elements in (7.21) and (7.22) corresponding to four of the state variables are the same, since the state variables are the same. But the value for the element X_4 in (7.21) and V in (7.22) are completely different since they physically represent different quantities.

7.3 FINDING EIGENVALUES AND EIGENVECTORS

The basic eigenvalue/eigenvector problem is to find the values of λ and $\mathbf{Y}$ that satisfy (7.11). This is a very difficult problem that has been solved only recently using sophisticated computer algorithms. The basic approach is as follows. Equation (7.11) can be written in the following manner:

$$\begin{bmatrix} (a_{11}-\lambda) & a_{12} & a_{13} & a_{14} \\ a_{21} & (a_{22}-\lambda) & a_{23} & a_{24} \\ a_{31} & a_{32} & (a_{33}-\lambda) & a_{34} \\ a_{41} & a_{42} & a_{43} & (a_{44}-\lambda) \end{bmatrix} \begin{bmatrix} X_1 \\ X_2 \\ X_3 \\ X_4 \end{bmatrix} = 0. \tag{7.23}$$

The only condition under which there can be a nonzero solution for $\mathbf{Y}$ is that the determinant of the matrix $(\mathbf{A} - \lambda\,\mathbf{F})$ be zero. The first part of the solution is to find the values of the λ's, the eigenvalues, that will make this be so. This determinant can be expanded into a polynomial

$$\lambda^4 + c_3\lambda^3 + c_2\lambda^2 + c_1\lambda + c_0 = 0. \tag{7.24}$$

However, there are neither satisfactory algorithms for going from (7.23) to (7.24) nor for factoring the resulting polynomial when the system is of a large size. But factoring can be used for low order systems, and this technique was used in Section 2.3 to establish the terms of the Park's transformation matrix.

The better approach is to realize that if the original matrix was either triangular or diagonal, the diagonal elements would be the eigenvalues. Also, it can be shown that if the **A** matrix is changed by a similarity transformation

$$\mathbf{B} = \mathbf{T}^{-1}\mathbf{A}\mathbf{T} \tag{7.25}$$

the matrix **B** will have the same eigenvalues as the matrix **A**. The trick is to find the transformation **T** such that **B** is triangular. The method to be used in this work is the so-called QR transform, an elegant but complex algorithm, which is described in [3]. Once the eigenvalues are found, the eigenvectors can be found by solving (7.23). If the λ's from the first step are numerically exact, this is a straightforward problem. However, the eigenvalue computation is usually slightly in error, due to round-off difficulties. To counter this, the eigenvectors are found using an inverse iteration method that is described in [4].

7.4 REFERENCES FOR CHAPTER 7

1. Jay, Frank, Ed., **IEEE Dictionary of Electrical and Electronic Terms, Second Edition**, IEEE Standard 100-1977, John Wiley and Sons, Interscience, New York, 1977.

2. Ogata, K., **State Space Analysis of Control Systems**, Prentice-Hall, 1967.

3. Martin, R. S., G. Peters, and J. H. Wilkinson, "The QR Algorithm for Real Hessenberg Matrices," Contribution II/14 in the **Handbook for Automatic Computation: Volume II, Linear Algebra**, Springer-Verlag, 1971.

4. Van Ness, J. E.. "Inverse Iteration Method for Finding Eigenvectors," **IEEE Trans. on Automatic Control**, v. AC-14, n. 1, Feb. 1969, p. 63-66.

CHAPTER 8

SSR EIGENVALUE ANALYSIS

The objective of detailed mathematical modeling and careful field testing for model parameters is to be able to perform accurate mathematical analysis. The goal of the analysis may be to determine the existence of a resonant condition, to study the effect of a change in control parameters on the damping of an oscillation, or to examine many other conditions. Several different types of SSR analysis have been used and are described in the literature. In this chapter, we concentrate only on eigenvalue analysis. This type of analysis is fundamental. It gives very important information regarding both the natural frequencies of oscillation of a system and the damping of each frequency. It is relatively easy, therefore, to determine those torsional frequencies that are not damped, and would therefore result in growing oscillations and almost certain damage to the affected turbine-generator shaft.

This chapter presents three solved SSR problems that are referred to here as "benchmark" cases. Two of the benchmark cases have been published by the IEEE as an aid to persons who are engaged in program preparation. These cases provide relatively simple problems for which the complete solution is known exactly. The third benchmark case, called the "CORPALS Benchmark," is a larger test case that is more typical of problems solved in industry. All of these problems illustrate the application of the theory presented in Chapters 2 through 7, and demonstrate the analysis of the SSR problem using eigenvalue computation. It is believed that, by studying the input data preparation for these three cases, the engineer will be able to see the scope of work that must be performed in order to perform eigenvalue analysis. These examples will also show computed results, and will illustrate the many different modes of oscillation that occur naturally in a power system. Most of these oscillatory modes are very well damped, but the troublesome modes are clearly identified. Moreover, the factors that contribute to these modes may be analyzed by computing the eigenvectors.

8.1 THE IEEE FIRST BENCHMARK MODEL

The IEEE First Benchmark Model (FBM) was created by the IEEE Working Group on Subsynchronous Resonance in 1977 for use in "computer program comparison and development." This small system is described by realistic parameters and provides a useful test bed for the SSR analytical methods. Here, the system is solved using the eigenvalue technique.

8.1.1 The FBM Network Model

The FBM network consists of a single series-capacitor compensated transmission line connecting a synchronous generator to a large system [1]. The network is shown in Figure 8.1. The network parameters are given in Table 8.1 except for the value of the series capacitor, which is usually varied in simulation studies.

The beauty of the FBM is its simplicity. There is only one interaction that can exist from this system, and that is the interaction between the machine and the simple one-line network.

In the simulations discussed in this section, we shall use a value of capacitive reactance of 0.35 per unit. This gives the total system impedance of

$$\begin{aligned}\mathbf{Z} &= R + jX_L - jX_C \\ &= 0.02 + j0.60 - j0.35 = 0.02 + j0.25 \text{ per unit.}\end{aligned} \tag{8.1}$$

The value of 0.35 capacitive reactance corresponds to 70% compensation of the 0.50 inductive reactance of the transmission line, and is a practical upper limit for series compensation of long transmission lines.

Using the given value of capacitance, we may compute the resonant frequency of the transmission line as follows.

$$\begin{aligned}\omega_o &= \frac{1}{\sqrt{LC}} = \frac{1}{\sqrt{(0.70)/(0.35)}} \\ &= 0.7071 \text{ per unit} = 266.57 \text{ rad/s.}\end{aligned} \tag{8.2}$$

The zero sequence impedances shown in Table 8.1 are not required for the examples computed here, but are included for the use of anyone interested in further analysis.

This corresponds to a frequency in hertz of

$$f_o = 42.426 \text{ Hz} \tag{8.3}$$

which is clearly in the range of frequencies that may give rise to a subsynchronous resonance with the turbine-generator shaft.

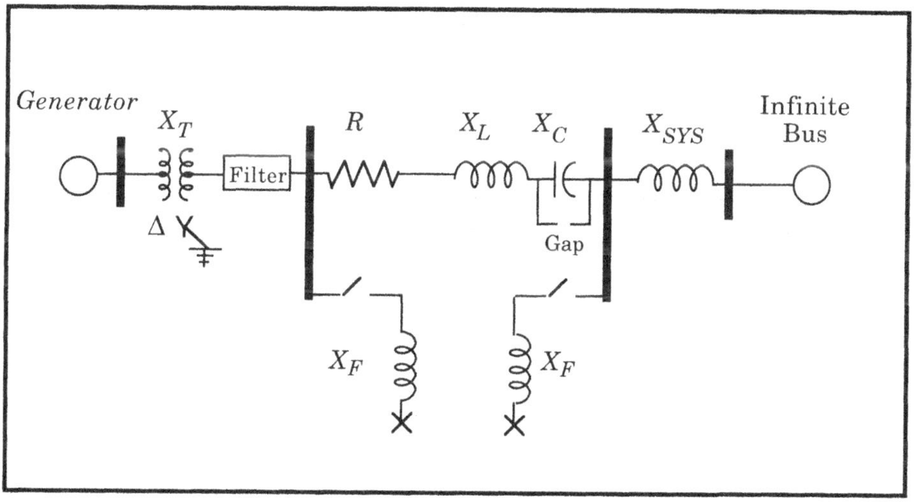

Figure 8.1 The First Benchmark Model Network [1].

Table 8.1 Network Impedances in Per Unit on the Generator MVA Base (892.4 MVA)

Parameter	Positive Sequence	Zero Sequence
R	0.02	0.50
X_T	0.14	0.14
X_L	0.50	1.56
X_{SYS}	0.06	0.06

In terms of the eigenvalue computation, we can expect eigenvalues with complex pairs in the neighborhood of

$$\omega = 2\pi f_B \pm \omega_o = 120\pi \pm 266.573 = 643.564 \text{ and } 110.418 \text{ rad/ s.} \tag{8.4}$$

8.1.2 The FBM Synchronous Generator Model

The generator model presented in the FBM specifications has two rotor circuits in each axis, exactly like the model presented in Chapter 2. The parameters of the FBM generator are taken from [1] and are listed in Table 8.2. All inductances are in per unit on the generator base and all time constants are in seconds.

Table 8.2 Synchronous Machine Parameters of the IEEE First Benchmark Model

Symbol	Inductances in per unit	Symbol	Time Constants in seconds
ℓ_a	0.130	τ'_{do}	4.300
L_d	1.790	τ''_{do}	0.032
L'_d	0.169	τ'_{qo}	0.850
L''_d	0.135	τ''_{qo}	0.050
L_q	1.710		
L'_q	0.228		
L''_q	0.200		

From the tabulated open circuit time constants we may derive the short circuit time constants using the formulas of Table 2.3, as follows:

$$\begin{aligned} \tau'_d &= 0.40598 \text{ per unit} \\ \tau''_d &= 0.02556 \text{ per unit} \\ \tau'_q &= 0.11333 \text{ per unit} \\ \tau''_q &= 0.04386 \text{ per unit.} \end{aligned} \tag{8.5}$$

8.1.3 The FBM Shaft Model

The turbine-generator shaft in the FBM is shown in Figure 8.2. This shaft model is typical of large turbine-generator shaft arrangements, where several turbine sections are modeled separately, as shown. The data are presented in [1] in the form of inertia constants, spring constants of the shaft sections connecting the inertias, and the modal dampings.

The inertia constants and spring constants for the shaft model shown in Figure 8.2 are given in Table 8.3.

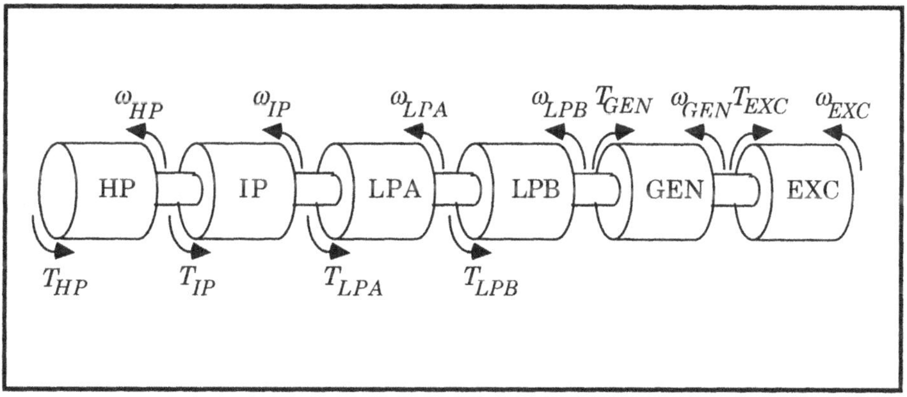

Figure 8.2 Turbine-Generator Shaft Model for the FBM

Table 8.3 Shaft Inertias and Spring Constants for the First Benchmark Model in Per Unit on the Machine Base [1]

Inertia	Inertia Constant H in s	Shaft Section	Spring Constant K in pu T/rad
HP Turbine	0.092897		
		HP-IP	19.303
IP Turbine	0.155589		
		IP-LPA	34.929
LPA Turbine	0.858670		
		LPA-LPB	52.038
LPB Turbine	0.884215		
		LPB-Gen	70.858
Generator	0.868495		
		Gen-Exc	2.82
Exciter	0.0342165		

The individual mass dashpot dampings are not available from the manufacturers, and such dampings are not provided as part of the FBM data. Instead, the modal dampings are given, which is the usual practice in the industry. These modal dampings are tabulated in terms of the decrement factors, σ_n, as discussed in Chapter 7. The no-load decrement factors for the first four modes are shown in Table 8.4.

The computed results of SSR/EIGEN [3], using the data given above, are plotted in Figure 8.3 and are tabulated in Table 8.5.

Table 8.4 No-Load Modal Decrement Factors for the First Benchmark Model [1]

Mode	H_n	f_n	σ_n
	in s	in Hz	in s^{-1}
1	2.70	15.71	0.050
2	27.80	20.21	0.110
3	6.92	25.55	0.028
4	3.92	32.28	0.028

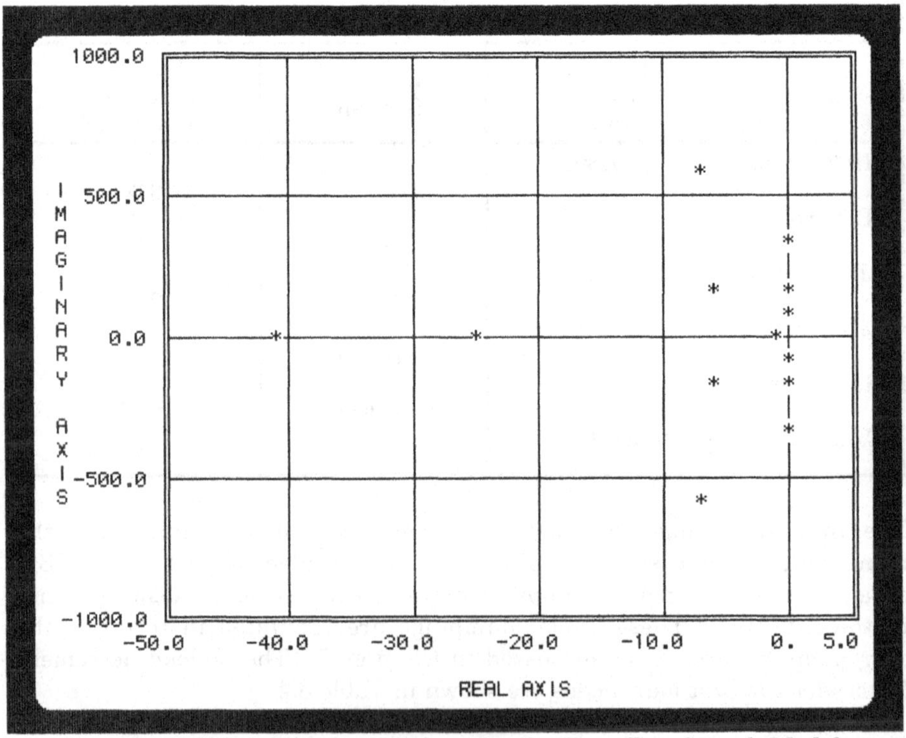

Figure 8.3 Plot of the Eigenvalues for the First Benchmark Model

Table 8.5 Computed Eigenvalues for the First Benchmark Model

Eigenvalue Number	Real Part, s^{-1}	Imaginary Part, rad/s	Imaginary Part, Hz
1,2	+0.07854636	±127.15560200	±20.2374426
3,4	+0.07818368	±99.70883066	±15.86915327
5,6	+0.04089805	±160.38986053	±25.52683912
7,8	+0.00232994	±202.86306822	±32.28666008
9,10	−0.00000048	±298.17672924	±47.45630037
11	−0.77576318		
12	−0.94796049		
13,14	−1.21804111	±10.59514740	±96.61615878
15,16	−5.54108044	±136.97740321	±21.80063081
17,18	−6.80964255	±616.53245850	±98.12275595
19	−25.41118956		
20	−41.29551248		

Recall, from (8.4), that the predicted network resonance frequencies of 643 and 110 radians per second were observed to be in the frequency neighborhood that might lead to resonance with the turbine generator shaft. Two of the unstable eigenvalues are near 110 radians per second, one at 99.7, and another at 128.2. This is surely due to the critical value of the capacitance chosen.

The system is twentieth order and there are ten eigenvalues with positive real parts. This is due to the value of series capacitance chosen. Repeating the calculation with other values of capacitance will change the eigenvalues and can result in a stable system.

8.2 THE IEEE SECOND BENCHMARK MODEL

The IEEE Second Benchmark Model (SBM) is similar to the First Benchmark Model in that the systems are small and easy to implement in a computer simulation. The major difference is that the Second Benchmark Model has two systems, one of which has two generators [2]. Moreover, the IEEE Second Benchmark Model deals with the "parallel resonance" problem and the interaction between turbine-generators that have a common mode of oscillation.

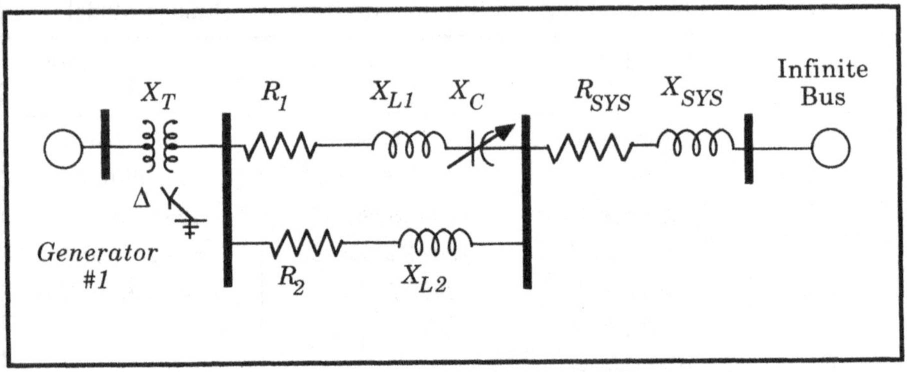

Figure 8.4 Second Benchmark Model System #1

Table 8.6 System #1 Network Impedances in Per Unit Based on 100 MVA and 500 kV

Parameter	Positive Sequence	Zero Sequence
R_T	0.0002	0.0002
X_T	0.0200	0.0200
R_1	0.0074	0.0220
X_{L1}	0.0800	0.2400
R_2	0.0067	0.0186
X_{L2}	0.0739	0.2100
R_{SYS}	0.0014	0.0014
X_{SYS}	0.0300	0.0300

8.2.1 Second Benchmark Model--System #1

The first system provided in the Second Benchmark Model paper [2] is shown in Figure 8.4 and is referred to herein as System #1.

The value of capacitive reactance is not specified explicitly, but is specified as a variable, taking on values of from 10% to 90% of series inductive reactance of the same line. This system is designed especially for the study of negative damping due to self excitation, which may be computed as a function of the amount of series compensation.

The data for the SBM are given in Table 8.6. All data are given on a 100 MVA base and the line impedances are on a 500 kV base.

Note that the transmission lines are similar, but are not identical. Also, note that the line with series compensation is the line designated #1 in Figure 8.4. The system to which the generator is connected is fairly strong, having low Thevenin impedance, and the inertia of this system is infinite.

System #1 of the SBM paper was designed for the study of self excitation of the unit as a function of the series compensation, and for the study of torque amplification in the first subsynchronous mode of oscillation [2].

8.2.2 Second Benchmark Model--System#2

Model #2 from the Second Benchmark Model paper [2] is shown in Figure 8.5. This system has two generators that have a common torsional mode of oscillation. The two generators are connected to a single series-compensated transmission line, and through this line to a very large system. The data for this system are given in Table 8.7. Note that the two transformers are designated #1 and #2 in agreement with the numbering of the two generators. We also note that Generator #1 and its step up transformer are identical in each of the two systems. The data for both generators is given in the next section.

As with System #1, the capacitive reactance is not assigned a specific value as the amount of compensation is the subject of study. Normally, the capacitive reactance will be varied in the range of 10 to 90% of the line inductive reactance.

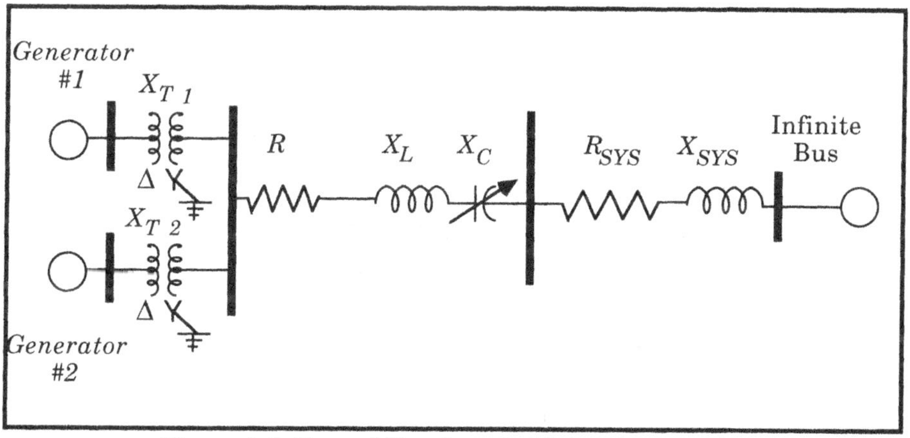

Figure 8.5 Second Benchmark Model System #2

Table 8.7 System #2 Network Impedances in Per Unit Based on 100 MVA and 500 kV

Parameter	Positive Sequence	Zero Sequence
R_{T1}	0.0002	0.0002
X_{T1}	0.0200	0.0200
R_{T2}	0.0004	0.0004
X_{T2}	0.0400	0.0400
R	0.0052	0.0120
X_L	0.0540	0.1200
R_{SYS}	0.0014	0.0014
X_{SYS}	0.0300	0.0300

8.2.3 SBM Generator, Circuit, and Shaft Data

The generator circuit and time constant data for the two units are given in Table 8.8. The shaft spring-mass models for the two units are shown in Figure 8.6.

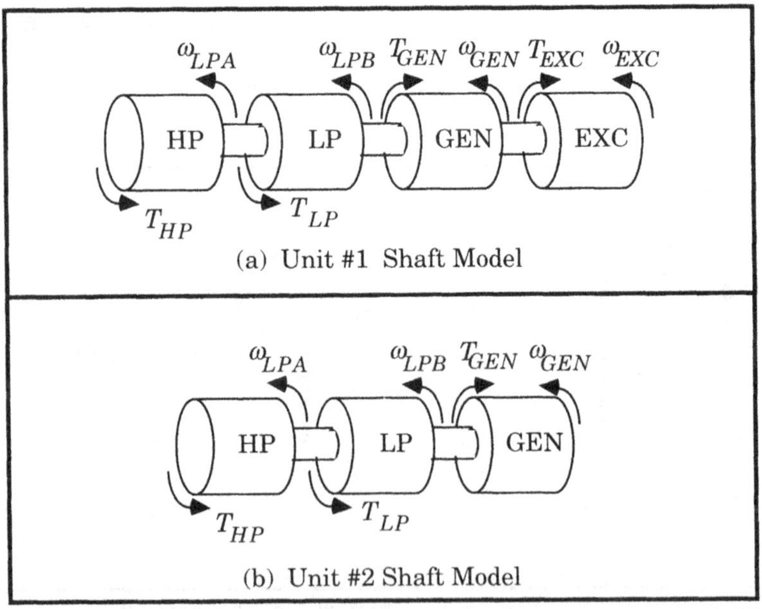

Figure 8.6 Spring Mass Shaft Models for the Second Benchmark Model

Table 8.8 Typical Synchronous Machine Parameters in the d-q Reference Frame [2,3]

Symbol	Inductances in Per Unit	
	Unit #1	Unit #2
r_a	0.0045	0.0045
ℓ_a	0.140	0.120
L_d	1.650	1.540
L'_d	0.250	0.230
L''_d	0.200	0.180
L_q	1.590	1.500
L'_q	0.460	0.420
L''_q	0.200	0.180
Symbol	**Time Constants in Seconds**	
	Unit #1	**Unit #2**
τ'_{do}	4.500	3.700
τ''_{do}	0.040	0.040
τ'_{qo}	0.550	0.430
τ''_{qo}	0.090	0.060
Rating	**Generator Ratings**	
	Unit #1	**Unit #2**
Rated MVA	600.0	700.0
Rated kV	22.0	22.0

Note that Unit #1 is modeled as a four mass shaft and that Unit #2 is modeled as a three mass shaft. The data for these spring-mass shaft models are given in English units, as is often the case in North American practice.

The shaft spring-mass data for Unit #1 are given in Table 8.9. Note the use of English units for all quantities. Dampings were chosen for these benchmark cases to be proportional to the inertias so that each mode has the same torsional damping in radians per second. Therefore, the modal dampings are directly related to the viscous dampings of the elements.

Table 8.9 Shaft Inertias, Dampings, and Spring Constants for the Second Benchmark Model Unit #1 [2]				
Mass	Inertia $lbm\text{-}ft^2$	Damping lbf-ft-sec/rad	Shaft Section	Spring Constant lbf-ft/rad
Exc	1,383	4.3	Exc-Gen	4.39×10^6
Gen	176,204	547.9	LP-Gen	97.97×10^6
LP	310,729	966.2	HP-LP	50.12×10^6
HP	49,912	155.2	...	...

Two other kinds of data normally provided by turbine-generator manufacturers are the rotor mode shapes and the computed model frequencies, decrement factors, and inertia constants.

The rotor mode shapes are given in Table 8.10 and are normalized with the generator mode taking on a value of unity. Finally, the computed modal quantities for Unit #1 are given in Table 8.11.

Table 8.10 Rotor Mode Shapes for Second Benchmark Unit #1			
Rotor	Mode 1	Mode 2	Mode 3
Exc	1.307	1.683	– 102.6000
Gen	1.000	1.000	1.0000
LP	– 0.354	– 1.345	– 0.1180
HP	– 1.365	4.813	0.0544

Table 8.11 Computed Modal Quantities for Second Benchmark Unit #1 [2]

Mode	f_n	σ_n	H_n
	Hz	rad/s	seconds
1	24.65	0.05	1.55
2	32.39	0.05	9.39
3	51.10	0.05	74.80

The shaft torsional data for generating Unit #2 are given in Table 8.12.

Table 8.12 Shaft Inertias, Dampings, and Spring Constants for the Second Benchmark Model Unit #2 [2]

Mass	Inertia lbm-ft^2	Damping lbf-ft-sec/rad	Shaft Section	Spring Constant lbf-ft/rad
Gen	334,914	208.20	LP-Gen	156.1×10^6
LP	370,483	230.40	HP-LP	198.7×10^6
HP	109,922	68.38	...	...

The Unit #2 rotor mode shapes are given in Table 8.13.

Table 8.13 Rotor Mode Shapes for Second Benchmark Unit #2

Rotor	Mode 1	Mode 2
Gen	1.000	1.00
LP	–0.601	– 4.33
HP	– 1.023	11.56

Finally, the computed modal quantities for Unit #2 are given in Table 8.14

Table 8.14 Computed Modal Quantities for Second Benchmark Unit #2 [2]			
Mode	f_n	σ_n	H_n
	Hz	rad/s	seconds
1	24.65	0.025	2.495
2	44.99	0.025	93.960

8.2.4 Computed Results for the Second Benchmark Models

The Second Benchmark Model has two sets of solutions, one for each of the defined systems. These solutions are given below, beginning for SBM System #1.

The computed results of the SSR/EIGEN computation are given in Table 8.15 and plotted in Figure 8.7.

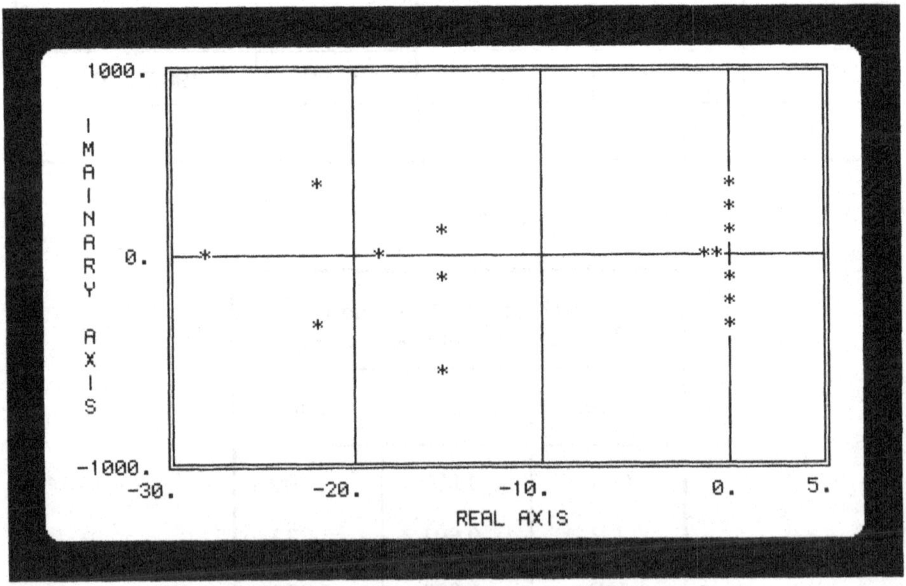

Figure 8.7 Plot of the SBM System #1 Eigenvalues

Table 8.15 Computed Eigenvalues for the 2nd Benchmark Model, Case 1

Eigenvalue Number	Real Part s^{-1}	Imaginary Part rad/s	Imaginary Part Hz
1,2	– 0.281,190,58	± 155.170,526,35	± 24.696,156,28
3,4	– 0.049,544,50	± 321.133,050,28	± 51.109,912,33
5,6	– 0.179,096,06	± 203.461,067,84	± 36.679,018,13
7	– 0.637,706,36		
8	– 1.207,327,28		
9,10	– 1.651,401,71	9.656,853,72	± 1.536,936,00
11,12	– 15.384,468,30	± 148.637,114,63	± 23.656,331,52
13,14	– 15.621,987,30	± 605.500,717,04	± 96.368,432,16
15	– 18.781,329,61		
16,17	– 21.767,731,96	± 376.908,072,02	± 59.986,782,75
18	– 27.951,255,03		

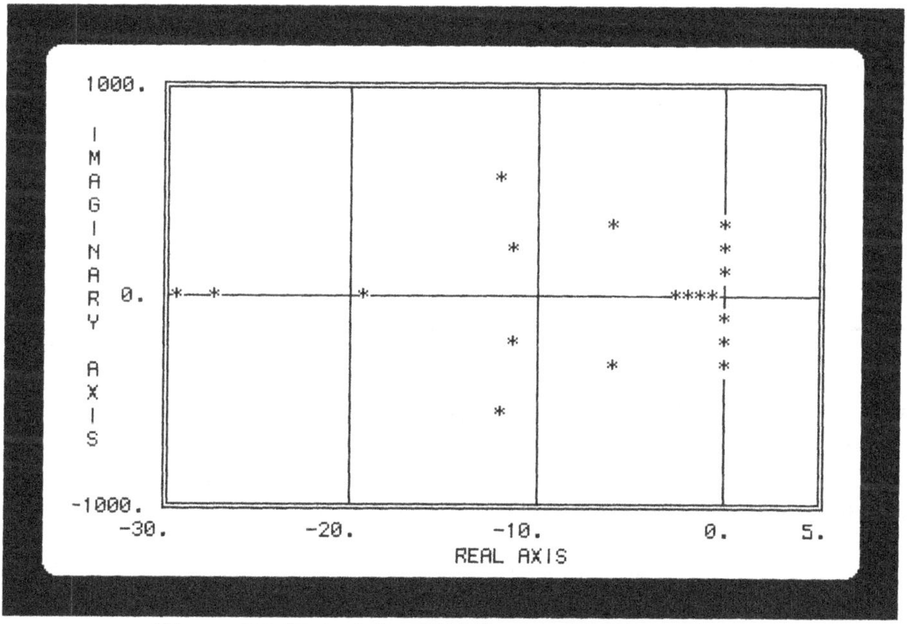

Figure 8.8 Plot of the SBM System #2 Eigenvalues

The computed results of the SSR/EIGEN computation are given in Table 8.16 and plotted in Figure 8.8.

Table 8.16 Computed Eigenvalues for the SBM, System #2

Eigenvalue Number	Real Part s^{-1}	Imaginary Part rad/s	Imaginary Part Hz
1,2	+ 0.310,146,11	± 155.658,710,28	± 24.773,853,16
3,4	+ 0.092,852,59	± 155.639,894,66	± 24.770,858,57
5,6	– 0.010,029,16	± 282.785,190,65	± 45.006,660,92
7,8	– 0.050,067,19	± 321.134,914,84	± 51.110,209,09
9,10	– 0.180,904,97	± 203.469,348,36	± 32.383,152,55
11	– 0.534,821,14		
12,13	– 0.971,083,62	± 6.926,607,29	± 1.102,403,79
14	– 0.984,581,26		
15	– 1.207,332,88		
16	– 1.704,824,31		
17,18	– 2.631,858,12	± 11.500,986,87	± 1.830,438,91
19,20	– 6.316,788,71	± 377.070,768,27	± 60.012,676,65
21,22	– 11.348,185,04	± 167.089,906,76	± 26.593,184,59
23,24	– 12.085,759,72	± 586.318,569,87	± 93.315,498,61
24	– 19.104,747,10		
26	– 27.412,035,27		
27	– 27.575,178,66		
28	– 29.192,896,44		

8.3 THE CORPALS BENCHMARK MODEL

The CORPALS Benchmark model is based on a system of more practical size for utility analysis than the IEEE benchmark models. This new benchmark, which is shown in Figure 8.9, has the following size:

Number of Buses: 29
Number of Branches: 39
Number of Generators: 5.

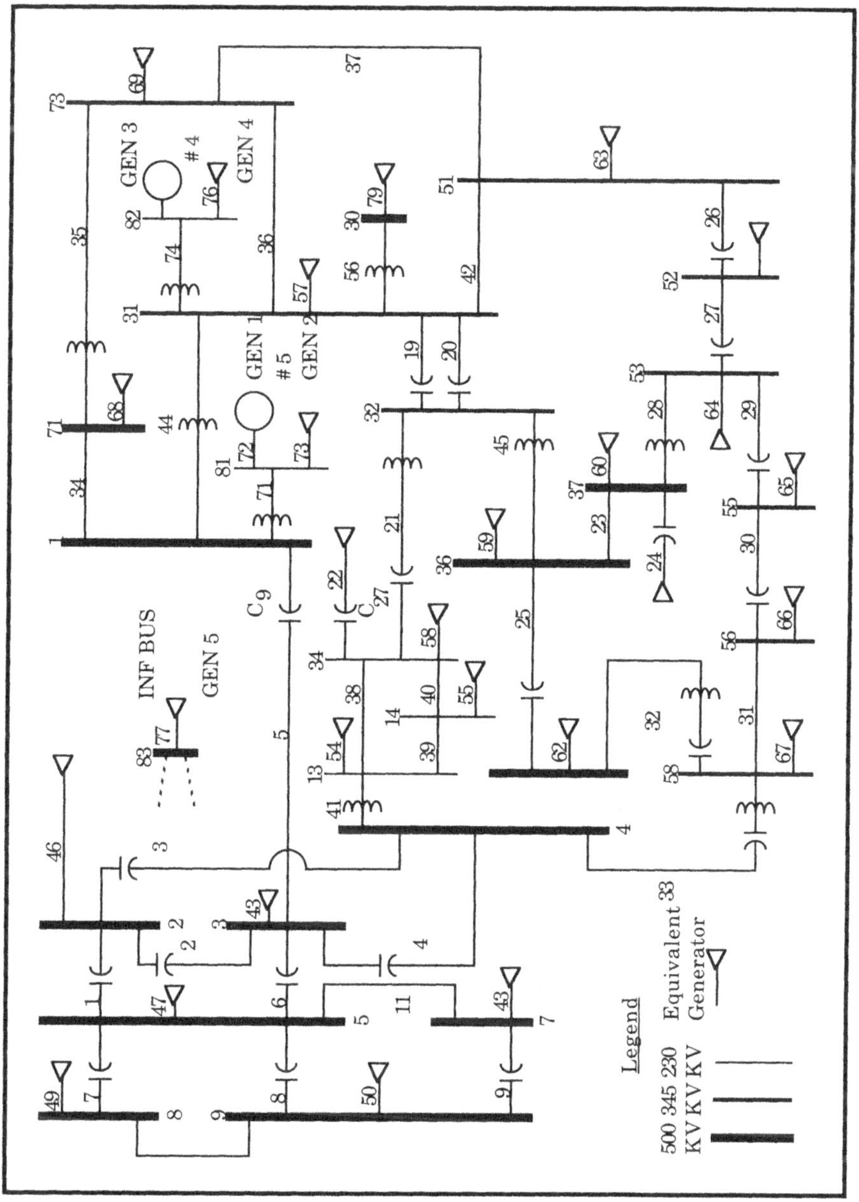

Figure 8.9 One Line Diagram of the CORPALS Benchmark Model

Table 8.17 CORPALS Benchmark Network Data in Per Unit on a 100 MVA Base

Branch Number	From Bus	To Bus	R	X_L	X_C
1	2	5	0.00250	0.06190	0.04330
2	2	3	0.00080	0.01800	0.01260
3	2	4	0.00240	0.05870	0.02350
5	1	3	0.00180	0.04200	0.01390
6	3	5	0.00210	0.04960	0.03470
7	5	8	0.00180	0.03910	0.01020
8	5	9	0.00135	0.04160	0.01080
9	7	9	0.00134	0.04120	0.01080
10	8	9	0.00020	0.00410	0.00000
11	5	7	0.00060	0.01420	0.00000
19	31	32	0.00460	0.04485	0.02245
21	32	34	0.00391	0.04220	0.01500
22	34	83	0.00950	0.11970	0.04310
38	13	34	0.00078	0.00780	0.00000
39	13	14	0.00399	0.03230	0.00000
40	14	34	0.00194	0.01600	0.00000
41	4	13	0.00004	0.00270	0.00000
43	3	83	0.01000	1.00000	0.00000
44	1	31	0.00019	0.01180	0.00000
45	32	36	0.00012	0.00760	0.00000
46	2	83	0.00000	0.01320	0.00000
47	5	83	0.00164	0.03333	0.00000
48	7	83	0.00000	0.02080	0.00000
49	8	83	0.00150	0.02680	0.00000
50	9	83	0.00023	0.00664	0.00000
54	13	83	0.00122	0.06304	0.00000
55	14	83	0.00411	0.08119	0.00000
56	30	31	0.00012	0.00770	0.00000
57	31	83	0.00800	0.09680	0.00000
58	34	83	0.00163	0.05487	0.00000
71	1	81	0.00036	0.02153	0.00000
72	0	81	0.00000	0.00000	0.00000
73	0	81	0.00000	0.00000	0.00000
74	31	82	0.00032	0.01883	0.00000
75	0	82	0.00000	0.00000	0.00000
76	0	82	0.00000	0.00000	0.00000
77	0	83	0.00000	0.00000	0.00000
78	30	83	0.00200	0.07000	0.00000
79	32	83	0.01070	0.22750	0.00000

The system consists of an interconnection of 500, 345, and 230 kV transmission lines, such as might be found for a typical power system. The system is connected to an infinite bus at node 83. An examination of the line data will show that several nodes are connected to this bus. The triangles in Figure 8.9 represent equivalent generators or interconnections. In an actual system, the engineer will probably represent one or two units in detail, but may move the detailed representation, as required, in order to study all machines of interest.

8.3.1 The CORPALS Network Model

The network data for the system is shown in Table 8.17. Note that not all lines have series compensation. Of the 39 branches in the model, only 11 are compensated transmission lines. The transmission system in this model represents three different levels of transmission voltage, as noted above. Series compensation is included in the longer 500 and 345 kV circuits. These lines are noted in Figure 8.9 by capacitor symbols. The per cent series compensation values are in the range of 25 to 70%. These are practical values for transmission lines in service in the Western United States, with 70% compensation being a practical upper limit for lines in this region.

Clearly, there is quite a lot of data for a system of this size and data entry into the eigenvalue program is time consuming. Note, however, that this needs to be done only once in any practical system. After the initial data entry, the data file may be saved and simply modified after that to simulate alternative system conditions. The data entry problem is also made simpler by representing some of the generators by equivalents. This leaves out some of the eigenvalues, but permits the user to concentrate on those eigenvalues that are related to the machine under study.

8.3.2 The CORPALS Machine Models

The synchronous machine parameters are given in Table 8.18. Only two of the generating units are represented in detail. Although it is possible to represent many of the units in detail, this is usually not necessary since the engineer is interested in the performance of only one unit at a time. Moreover, representing many units in detail tends to give a large number of machine eigenvalues that are close together, thereby making it more difficult to identify those of interest. Therefore, in many cases, it is practical to represent most of the machines by classical models, with only those machines of direct interest being represented in detail. The detailed representation used here is based on two rotor branches in the d- and q-axes of the rotor.

Table 8.18 Synchronous Machine Parameters for the CORPALS Benchmark Model

Symbol	Inductances in Per Unit				
	Unit #1	Unit #2	Unit #3	Unit #4	Unit #5
Type	2	1	2	1	1
Name	GGG1	GGG2	GGG3	GGG4	GGG5
r_a	0.0027	0.000	0.0027	0.000	0.000
ℓ_a	0.160	0.205	0.160	0.205	0.000
L_d	1.800		1.800		
L_d'	0.249		0.249		
L_d''	0.201		0.201		
L_q	1.720		1.720		
L_q'	0.321		0.321		
L_q''	0.201		0.201		
Symbol	**Time Constants in Seconds**				
	Unit #1	Unit #2	Unit #3	Unit #4	Unit #5
τ_{do}'	1.2100		1.2100		
τ_{do}''	0.0095		0.0095		
τ_{qo}'	0.1720		0.1720		
τ_{qo}''	0.0049		0.0049		
Rating	**Other Parameters**				
	Unit #1	Unit #2	Unit #3	Unit #4	Unit #5
MVA	483.0	426.0	483.0	426.0	892.0
Taps	1.075	1.075	1.077	1.077	1.00

8.3.3 The CORPALS Eigenvalues

The computed eigenvalues for the CORPALS benchmark model are plotted in Figures 8.10 and are listed in Tables 8.19(a) and (b).. There are 88 eigenvalues for this system, 40 complex pairs, and eight real. For the case run here, using the data tabulated, one of the eigenvalues has a positive real part. Six of them are very near the imaginary axis, however. This is

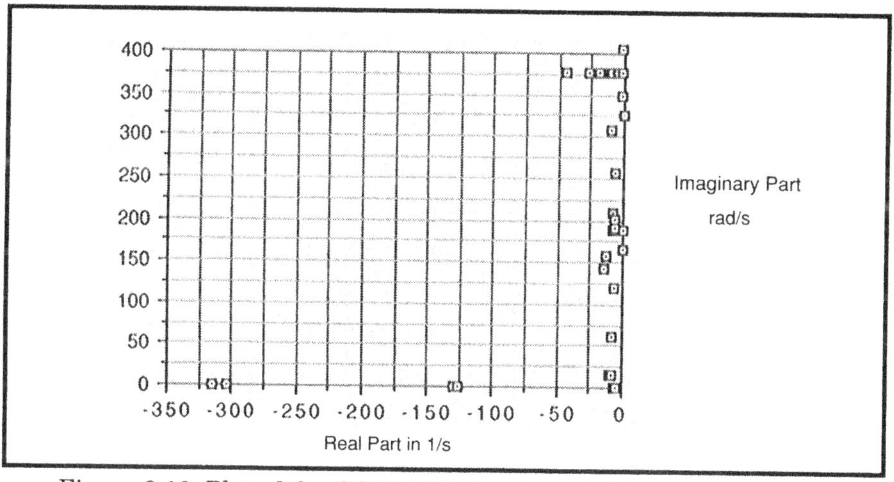

Figure 8.10 Plot of the CORPALS Benchmark Model Eigenvalues

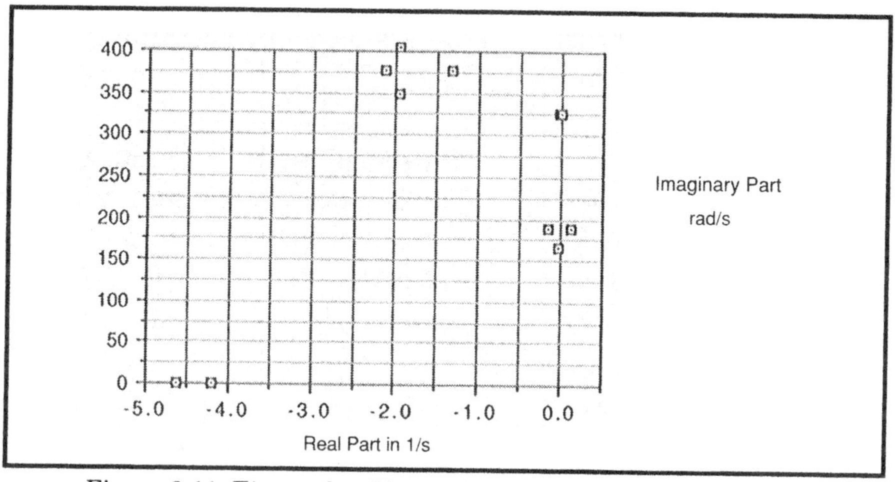

Figure 8.11 Eigenvalue Plot with Expanded Horizontal Scale

shown more clearly in Figure 8.11, which is plotted to a different scale. This clearly shows the eigenvalues that are close to the imaginary axis. Only one of the eigenvalues has a positive real part for the system case represented. The eigenvalues shown in Table 8.19 are listed in descending order of the eigenvalue real part. It from this table, and from Figure 8.11, it is clear that several eigenvalues are very poorly damped.

Table 8.19(a) Computed Eigenvalues for CORPALS Benchmark Model

Eigenvalue Number	Real Part s^{-1}	Imaginary Part rad/s	Imaginary Part Hz
1,2	+ 0.131,595,03	± 188.718,471,55	± 30.035,477,59
3,4	− 0.009,704,28	± 326.986,686,32	± 52.041,547,45
5,6	− 0.015,055,23	± 327.004,391,40	± 52.044,365,30
7,8	− 0.017,645,18	± 164.429,915,69	± 26.169,833,86
9,10	− 0.027,615,86	± 164.432,159,16	± 26.170,190,92
11,12	− 0.156,887,16	± 188.678,740,78	± 30.029,154,24
13,14	− 1.330,997,35	± 376.966,979,09	± 59.996,158,09
15,16	− 1.956,198,67	± 349.217,902,47	± 55.579,755,38
17,18	− 1.958,847,00	± 404.764,128,25	± 64.420,211,72
19,20	− 2.147,386,14	± 376.858,232,76	± 59.978,850,59
21	− 4.207,426,40		
22	− 4.630,864,95		
23,24	− 5.260,180,59	± 376.508,839,33	± 59.923,242,89
25,26	− 6.288,104,62	± 256.441,245,38	± 40.813,891,80
27,28	− 6.289,447,03	± 497.541,193,13	± 79.186,140,27
29,30	− 6.637,808,03	± 553.321,386,38	± 88.063,833,75
31,32	− 6.637,811,59	± 200.660,727,04	± 31.936,146,59
33	− 6.686,961,81		
34	−6.809,366,37		
35,36	−7.096,280,61	±190.136,052,34	±30.261,092,58
37,38	−7.188,127,84	±119.023,912,41	±18.943,244,00
39,40	−7.208,815,51	±565.259,228,09	±89.963,800,26
41,42	−7.208,977,28	±188.723,664,72	±30.036,304,12
43,44	−7.255,950,22	±634.985,511,34	±101.061,082,90

Table 8.19(b) Computed Eigenvalues for CORPALS Benchmark Model			
Eigenvalue Number	Real Part s^{-1}	Imaginary Part rad/s	Imaginary Part Hz
45,46	– 7.674,179,76	± 209.054,523,39	± 33.272,060,76
47,48	– 7.687,445,15	± 544.968,398,64	± 86.734,414,47
49,50	– 7.860,111,26	± 61.763,261,81	± 9.829,928,42
51,52	– 7.860,111,26	± 692.218,938,21	± 110.170,065,70
53,54	– 8.167,051,98	± 14.773,761,12	± 2.351,317,11
55,56	– 8.288,391,35	± 564.038,989,00	± 89.769,593,20
57,58	– 8.842,770,32	± 376.819,394,75	± 59.972,669,32
59,60	– 9.117,396,48	± 15.167,362,40	± 2.413,960,70
61,62	– 9.528,829,13	± 307.766,210,79	± 48.982,513,77
63,64	– 9.546,804,43	± 446.214,668,96	± 71.017,270,24
65,66	– 10.643,735,44	± 376.991,100,00	± 59.999,997,07
67,68	– 11.995,396,38	± 157.392,546,81	± 25.049,801,83
69,70	–12.241,490,62	±596.733,638,50	±94.973,108,28
71,72	–12.379,482,18	±376.991,100,00	±59,999,997,07
73,74	–14.051.671,92	±142.857,795,70	±22.736,524,35
75,76	–14.108,131,68	±611.112,152,76	±97.261,519,89
77,78	–19.215,605,52	±376.991,100,00	±59,999,997,07
79,80	–19.383,734,44	±376,991,100,00	±59,999,997,07
81,82	–27.010,398,60	±376.849,928,90	±59,977,528,99
83,84	–45.106,756,97	±376,991,100,00	±59,999,997,07
85	–126.608,680,09		
86	–128.764,341,23		
87	–302.855,958,03		
88	314.041,264,97		

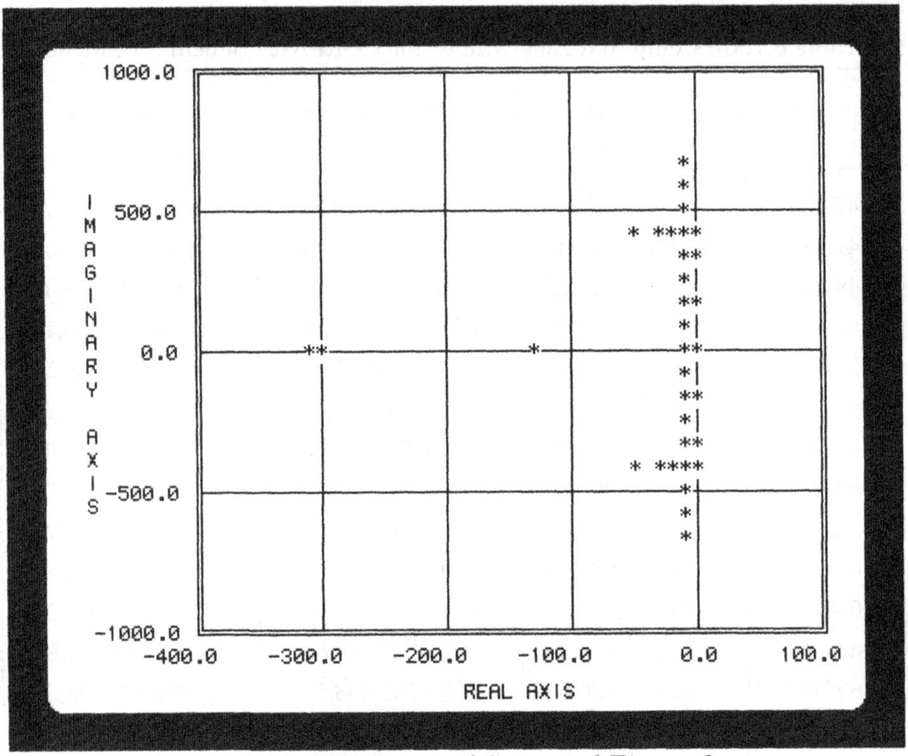

Figure 8.12 CRT Screen of Computed Eigenvalues

Another view of the plotted eigenvalues is shown in Figure 8.12. This is the CRT screen displayed by the computer program, which shows both the positive and negative frequencies. These screens are helpful for a quick check of the computation, but the scale is such that it is almost impossible to detect those eigenvalues with positive real parts.

8.4 AN EXAMPLE OF SSR EIGENVALUE ANALYSIS

We illustrate the use of SSR eigenvalue analysis by presenting an example of calculations on an actual power system. The system studied is the power plant and network associated with the Palo Verde Nuclear Generating Station (PVNGS), near Phoenix. The studies were performed by engineers at Arizona Public Service Company, including one of the authors of this book.

The Palo Verde Nuclear Generating Station consists of three identical 1270 MW generators owned by a consortium of utilities in the Southwestern

United States and operated by Arizona Public Service Company. Five 500 kV transmission lines connect the Palo Verde units to load centers in Phoenix, Los Angeles and San Diego areas as shown in Figure 8.13.

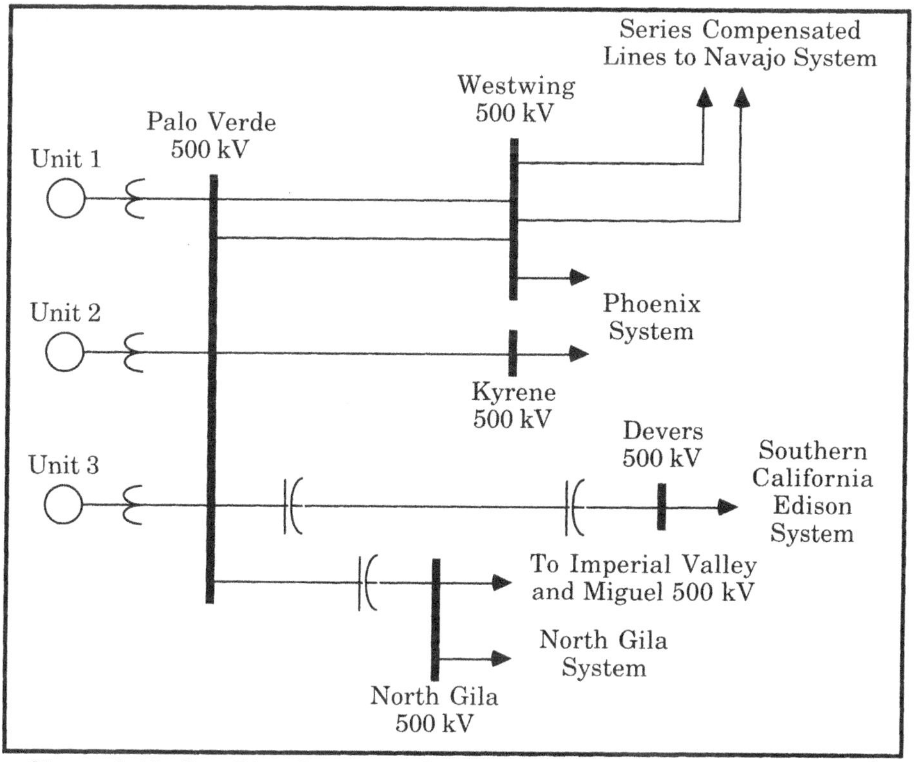

Figure 8.13 One Line Diagram of the Palo Verde Transmission System

Two of the lines shown in Figure 8.13 are series compensated. A thorough SSR analysis was conducted to evaluate the potential for SSR problems at the Palo Verde generating station. The analysis indicates that PVNGS is faced with potential SSR problems of the torsional interaction type. This type of problem is best evaluated by an eigenvalue analysis.

8.4.1 The Spring-Mass Model

Figure 8.14 shows the Palo Verde turbine-generator shaft model. The manufacturer has provided a 15 mass model of the turbine generator system. An alternate six-mode modal model was also available and was used for most of the studies.

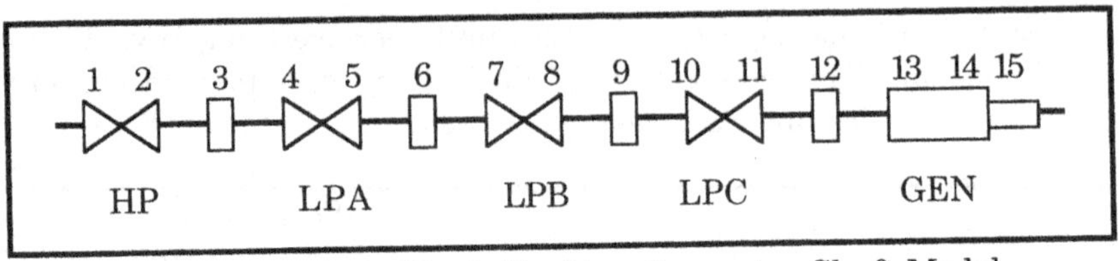

Figure 8.14 Palo Verde Turbine Generator Shaft Model

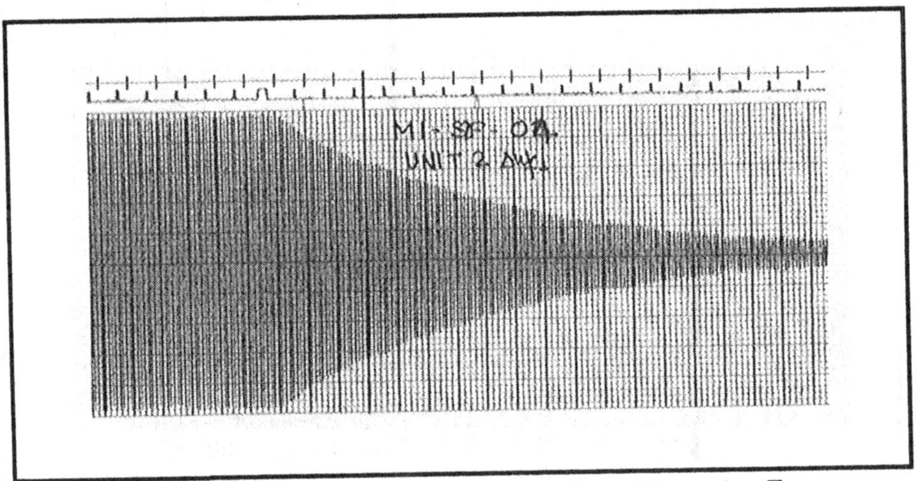

Figure 8.15 Typical Brush Chart Recording of Damping From Palo Verde SSR Tests

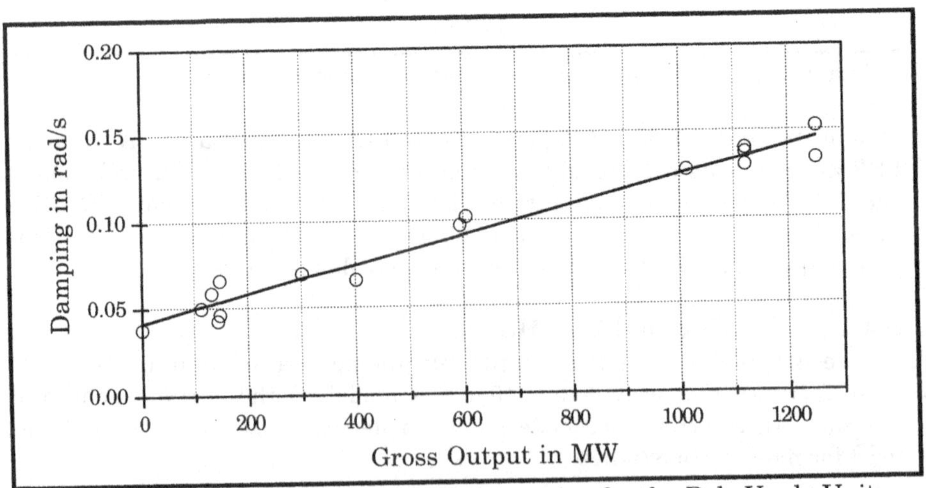

Figure 8.16 Typical Load Dependent Damping for the Palo Verde Units

Figures 8.15 show a typical test of mode 4 damping, where the gradual damping following the excitation of modal oscillation is illustrated. Figure 8.16 shows the results of several damping measurements that were taken at different values of machine loading. Note the almost linear increase in damping with load.

8.4.2 The System Eigenvalues

The measured damping factors are in the form of modal damping. Usually eigenvalue studies are conducted with zero damping represented for the turbine generator. A typical output from an eigenvalue program run is shown in Table 8.20. The output consists of one set of eigenvalues for each specific system condition. All eigenvalues are referred to the generator rotor side. Hence, the electrical system eigenvalues appear in the form of the 60 Hz complement. The eigenvalues are plotted in Figure 8.17.

Table 8.20 Typical Output From An Eigenvalue Analysis

Mode	Real Part rad/s	Imag Part rad/s	Imag Part Hz
4	0.02482867	146.14109894	23.2591
3	0.01359666	135.86212945	21.6231
	0.00007752	311.60004147	49.5927
2	-0.00521323	98.35874439	15.6541
1	-0.05876762	51.01832995	8.1198
	-0.32691083	0	0
	-0.53446581	0	0
	-0.69177605	6.94636329	1.1056
	-0.86602305	175.90263947	27.9958
	-7.73315173	117.59892687	18.7164
	-7.33152173	636.38331313	101.2840
	-7.80109906	577.87892388	91.9723
	-7.97460273	141.21203232	22.4746
	-8.02037455	612.76651856	97.5248
	-8.61372827	256.71130378	40.8569
	-8.63593329	497.26705814	79.1425
	-9.49697525	325.48695214	51.8029
	-9.50043339	428.49429850	68.1970
	-14.07485723	376.91695563	59.9882
	-15.13948573	0	0
	-27.88626710	376.99112000	60.0000
	-35.62657298	0	0

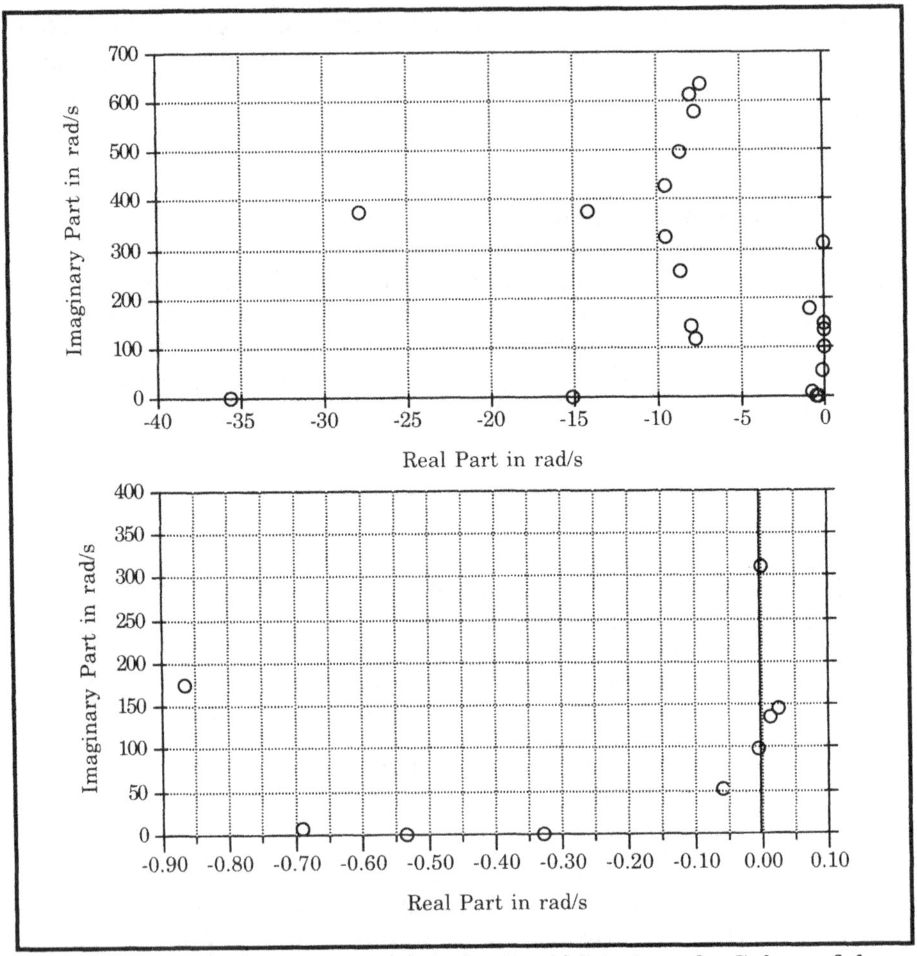

Figure 8.17 Computed Eigenvalues for PVNGS (top) and a Subset of the Eigenvalues Near the Imaginary Axis (bottom)

The eigenvalues associated with the mechanical system can be easily identified since their imaginary part (frequencies) are well known in advance or can be calculated by running the eigenvalue program without the electrical system represented. These frequencies do not change appreciably with various system operating conditions but the real parts of the eigenvalues will often change. For the Palo Verde units, there are four subsynchronous modes of concern. They are at 8.1, 15.7, 21.6 and 23.2 Hz and these are identified in Table 8.20.

Table 8.21 Analysis of Net Modal Damping					
Mode	Frequency rad/s	Hz	Eigenvalue Real Part (1)	Natural Damping (2)	Net Damping (3)
1	51.0	8.1	–0.059	0.030	0.089
2	98.4	15.7	–0.0052	0.045	0.050
3	135.9	21.6	+0.014	0.045	0.031
4	146.1	23.3	+0.025	0.030	0.005

The real part corresponding to these eigenvalues shown in Table 8.20, directly provides the damping, or negative damping, due to interaction with the system. If the real part is negative, it implies that interaction of the system is such that it improves the damping of the torsional mode. Such is the case for Modes 1 and 2 in Table 8.20. Thus, these modes would be stable even if they had zero damping. In reality, there is always some torsional damping.

8.4.3 Computation of Net Modal Damping

The net damping is the sum of the real part of the eigenvalue and the torsional damping. This calculation is shown in Table 8.21. If the real part of the eigenvalue is positive, it implies that torsional interaction produces negative damping in that mode. Such is the case for Modes 3 and 4 in Table 8.21. However, it does not necessarily mean that this mode is unstable, since the eigenvalue computation is made assuming zero damping. We must subtract the modal damping from the computed values to find the net damping. If the modal damping exceeds the real part of the eigenvalue, the mode will be damped in spite of the computed torsional interaction phenomenon. Such is the case for Modes 3 and 4 in Table 8.21.

Similar analysis must be made for other system operating conditions, including contingencies, to determine any potentially unstable operating system conditions. If unstable operating conditions are found, they must be avoided until an appropriate countermeasure are implemented. The eigenvalue analysis can be used to evaluate appropriateness of the countermeasure if adequate models are available.

8.5 REFERENCES FOR CHAPTER 8

1. IEEE Committee Report, "First Benchmark Model for Computer Simulation of Subsynchronous Resonance," **IEEE Trans.**, v. PAS-96, Sept/Oct 1977, p. 1565-1570.

2. IEEE Committee Report, "Second Benchmark Model for Computer Simulation of Subsynchronous Resonance," **IEEE Trans.,** v. PAS-104,n. 5, 1985, p. 1057-66.

3. **SSR/EIGEN, User's Manual for the Computation of Eigenvalues and Eigenvectors in Problems Related to Power System Subsynchronous Resonance**, Power Math Associates, Inc., Del Mar, CA, 1987.

Index

E

F

L

M

N

O

P

Q

R

S

U

V

W

ABOUT THE AUTHORS

P. M. Anderson (M'50–SM'56–F'81) received his B.S., M.S., and Ph.D. from Iowa State University in 1949, 1958, and 1961, respectively. He has been in the discipline of electric power systems for 40 years. Dr. Anderson has worked for utilities and as a consultant, but most of his career was as a university professor of electrical engineering with a special interest in power systems. His technical interests have been in power system mathematical modeling, system dynamic performance, reliability, and mathematical analysis. He is the author of two other graduate-level books on power systems.

B. L. Agrawal (S'74–M'74–SM'83) received his B.S. in electrical engineering from Birla Institute of Technology and Science, India, in 1970 and his M.S. and Ph.D. in control systems from the University of Arizona, Tucson, in 1972 and 1974, respectively. He joined the Arizona Public Service Company in 1974 and is now a Senior Consulting Engineer. His responsibilities include dynamic modeling and simulation of power system interaction with turbine generators, including the areas of power system stabilizer application, power system stability, subsynchronous resonance, and electric system transients. Dr. Agrawal is active in IEEE Working Groups related to system dynamic performance and in industry committees of the Western Systems Coordinating Council. He has authored a number of papers on power system dynamic performance.

J. E. Van Ness (S'52–M'52–SM'57–F'82) received his B.S. from Iowa State University in 1949 and his M.S. and Ph.D. from Northwestern University in 1951 and 1954, respectively. He has been on the faculty of Northwestern University since 1952. During this time his specialty has been the development of digital methods for analyzing power systems. His major contributions have been in load flow methods and eigenvalue analysis of power systems. Dr. Van Ness is the author of many technical papers on subjects related to power system analysis.

ABOUT THE AUTHORS

9 780780 353503

Made in the USA
Monee, IL
13 September 2026

62977541R00164